ASE Test Preparation

Automotive Technician Certification Series

Brakes (A5)
5th Edition

Australia • Brazil • Japan • Korea • Mexico • Singapore • Spain • United Kingdom • United States

ASE Test Preparation: Automotive Technician Certification Series, Brakes (A5), 5th Edition

Vice President, Technology and Trades Professional Business Unit: Gregory L. Clayton

Director, Professional Transportation Industry Training Solutions: Kristen L. Davis

Editorial Assistant: Danielle Filippone

Product Manager: Katie McGuire

Director of Marketing: Beth A. Lutz

Marketing Manager: Jennifer Barbic

Senior Production Director: Wendy Troeger

Production Manager: Sherondra Thedford

Content Project Management: PreMediaGlobal

Senior Art Director: Benjamin Gleeksman

Section Opener Image: Image Copyright Creations, 2012. Used under License from Shutterstock.com

© 2012, 2006, 2004, 2001 Delmar Cengage Learning

ALL RIGHTS RESERVED. No part of this work covered by the copyright herein may be reproduced, transmitted, stored or used in any form or by any means graphic, electronic, or mechanical, including but not limited to photocopying, recording, scanning, digitizing, taping, Web distribution, information networks, or information storage and retrieval systems, except as permitted under Section 107 or 108 of the 1976 United States Copyright Act, without the prior written permission of the publisher.

> For product information and technology assistance, contact us at
> **Cengage Learning Customer & Sales Support, 1-800-354-9706**
> For permission to use material from this text or product,
> submit all requests online at **www.cengage.com/permissions**
> Further permissions questions can be emailed to
> **permissionrequest@cengage.com**

ISBN-13: 978-1-111-12707-7

ISBN-10: 1-111-12707-7

Delmar
Executive Woods
5 Maxwell Drive
Clifton Park, NY 12065-2919
USA

Cengage Learning is a leading provider of customized learning solutions with office locations around the globe, including Singapore, the United Kingdom, Australia, Mexico, Brazil, and Japan. Locate your local office at **international.cengage.com/region**.

Cengage Learning products are represented in Canada by Nelson Education, Ltd.

To learn more about Delmar, visit **www.cengage.com/delmar**

Purchase any of our products at your local bookstore or at our preferred online store **www.cengagebrain.com**

Notice to the Reader

Publisher does not warrant or guarantee any of the products described herein or perform any independent analysis in connection with any of the product information contained herein. Publisher does not assume, and expressly disclaims, any obligation to obtain and include information other than that provided to it by the manufacturer. The reader is expressly warned to consider and adopt all safety precautions that might be indicated by the activities described herein and to avoid all potential hazards. By following the instructions contained herein, the reader willingly assumes all risks in connection with such instructions. The publisher makes no representations or warranties of any kind, including but not limited to, the warranties of fitness for particular purpose or merchantability, nor are any such representations implied with respect to the material set forth herein, and the publisher takes no responsibility with respect to such material. The publisher shall not be liable for any special, consequential, or exemplary damages resulting, in whole or part, from the readers' use of, or reliance upon, this material.

Printed in the United States of America
10 11 12 13 14 15 16   21 20 19 18 17

Table of Contents

Preface v

SECTION 1 The History and Purpose of ASE 1

SECTION 2 Overview and Introduction 2
    Exam Administration 2
    Understanding Test Question Basics 2
    Test-Taking Strategies 3
    Preparing for the Exam 4
    What to Expect During the Exam 5
    Testing Time. 5
    Understanding How Your Exam Is Scored 6

SECTION 3 Types of Questions on an ASE Exam 7
    Multiple-Choice/Direct Questions 7
    Completion Questions 8
    Technician A, Technician B Questions 8
    EXCEPT Questions 9
    LEAST LIKELY Questions 10
    Summary 10

SECTION 4 Task List Overview 11
    Introduction 11

SECTION 5 Sample Preparation Exams 46
    Introduction 46
    Preparation Exam 1 46

Preparation Exam 2 55
Preparation Exam 3 67
Preparation Exam 4 79
Preparation Exam 5 89
Preparation Exam 6 99

SECTION 6 Answer Keys and Explanations 109

Preparation Exam 1—Answer Key 109
Preparation Exam 1—Explanations 110
Preparation Exam 2—Answer Key 127
Preparation Exam 2—Explanations 127
Preparation Exam 3—Answer Key 148
Preparation Exam 3—Explanations 148
Preparation Exam 4—Answer Key 172
Preparation Exam 4—Explanations 172
Preparation Exam 5—Answer Key 190
Preparation Exam 5—Explanations 190
Preparation Exam 6—Answer Key 210
Preparation Exam 6—Explanations 210

SECTION 7 Appendices 229

Preparation Exam Answer Sheet Forms 229
Glossary 235

Preface

Delmar, a part of Cengage Learning, is very pleased that you have chosen to use our ASE Test Preparation Guide to help prepare yourself for the Brakes (A5) ASE certification examination. This guide is designed to help prepare you for your actual exam by providing you with an overview and introduction of the testing process, introducing you to the task list for the Brakes (A5) certification exam, giving you an understanding of what knowledge and skills you are expected to have in order to successfully perform the duties associated with each task area, and providing you with several preparation exams designed to emulate the live exam content in hopes of assessing your overall exam readiness.

If you have a basic working knowledge of the discipline you are testing for, you will find this book to be an excellent guide, helping you understand the "must know" items needed to successfully pass the ASE certification exam. This manual is not a textbook. Its objective is to prepare the individual who has the existing requisite experience and knowledge to attempt the challenge of the ASE certification process. This guide cannot replace the hands-on experience and theoretical knowledge required by ASE to master the vehicle repair technology associated with this exam. If you are unable to understand more than a few of the preparation questions and their corresponding explanations in this book, it could be that you require either more shop-floor experience or further study.

This book begins by providing an overview of, and introduction to, the testing process. This section outlines what we recommend you do to prepare, what to expect on the actual test day, and overall methodologies for your success. This section is followed by a detailed overview of the ASE task list to include explanations of the knowledge and skills you must possess to successfully answer questions related to each particular task. After the task list, we provide six sample preparation exams for you to use as a means of evaluating areas of understanding, as well as areas requiring improvement in order to successfully pass the ASE exam. Delmar is the first and only test preparation organization to provide so many unique preparation exams. We enhanced our guides to include this support as a means of providing you with the best preparation product available. Section 6 of this guide includes the answer keys for each preparation exam, along with the answer explanations for each question. Each answer explanation also contains a reference back to the related task or tasks that it assesses. This will provide you with a quick and easy method for referring back to the task list whenever needed. The last section of this book contains blank answer sheet forms you can use as you attempt each preparation exam, along with a glossary of terms.

OUR COMMITMENT TO EXCELLENCE

Thank you for choosing Delmar, Cengage Learning for your ASE test preparation needs. All of the writers, editors, and Delmar staff have worked very hard to make this test preparation guide second to none. We feel confident that you will find this guide easy to use and extremely beneficial as you prepare for your actual ASE exam.

Delmar, Cengage Learning has sought out the best subject matter experts in the country to help with the development of *ASE Test Preparation: Automotive Technician Certification Series, Brakes (A5),*

5th Edition. Preparation questions are authored and then reviewed by a group of certified subject-matter experts to ensure the highest level of quality and validity to our product.

If you have any questions concerning this guide or any guide in this series, please visit us on the web at **http://www.trainingbay.cengage.com**.

For web-based online test preparation for ASE certifications, please visit us on the web at **http://www.techniciantestprep.com** to learn more.

ABOUT THE SERIES ADVISOR

Mike Swaim has been an Automotive Technology Instructor at North Idaho College, Coeur d'Alene, Idaho, since 1978. He is an Automotive Service Excellence (ASE) Certified Master Technician since 1974 and holds a Lifetime Certification from Mobile Air Conditioning Society. He served as Series Advisor to all nine of the 2011 Automobile/Light Truck Certification Tests (A Series) of Delmar, Cengage ASE Test Preparation titles, and is the author of *ASE Test Preparation: Automotive Technician Certification Series, Undercar Specialist Designation (X1), 5th Edition*.

SECTION 1: The History and Purpose of ASE

ASE began as the National Institute for Automotive Service Excellence (NIASE). It was founded as a non-profit, independent entity in 1972 by a group of industry leaders with the single goal of providing a means for consumers to distinguish between incompetent and competent technicians. It accomplishes this goal through the testing and certification of repair and service professionals. Though it is still known as the National Institute for Automotive Service Excellence, it is now called "ASE" for short.

Today, ASE offers more than 40 certification exams in automotive, medium/heavy duty truck, collision repair and refinish, school bus, transit bus, parts specialist, automobile service consultant, and other industry-related areas. At this time, there are more than 385,000 professionals nationwide with current ASE certifications. These professionals are employed by new car and truck dealerships, independent repair facilities, fleets, service stations, franchised service facilities, and more.

ASE's certification exams are industry-driven and cover practically every on-highway vehicle service segment. The exams are designed to stress the knowledge of job-related skills. Certification consists of passing at least one exam and documenting two years of relevant work experience. To maintain certification, those with ASE credentials must be re-tested every five years.

While ASE certifications are a targeted means of acknowledging the skills and abilities of an individual technician, ASE also has a program designed to provide recognition for highly qualified repair, support, and parts businesses. The Blue Seal of Excellence Recognition Program allows businesses to showcase their technicians and their commitment to excellence. One of the requirements of becoming Blue Seal recognized is that the facility must have a minimum of 75 percent of their technicians ASE certified. Additional criteria apply, and program details can be found on the ASE website.

ASE recognized that educational programs serving the service and repair industry also needed a way to be recognized as having the faculty, facilities, and equipment to provide a quality education to students wanting to become service professionals. Through the combined efforts of ASE, industry, and education leaders, the non-profit organization entitled the National Automotive Technicians Education Foundation (NATEF) was created in 1983 to evaluate and recognize academic programs. Today more than 2,000 educational programs are NATEF certified.

For additional information about ASE, NATEF, or any of their programs, the following contact information can be used:

National Institute for Automotive Service Excellence (ASE)
101 Blue Seal Drive S.E.
Suite 101
Leesburg, VA 20175
Telephone: 703-669-6600
Fax: 703-669-6123
Website: **www.ase.com**

SECTION 2

Overview and Introduction

Participating in the National Institute for Automotive Service Excellence (ASE) voluntary certification program provides you with the opportunity to demonstrate you are a qualified and skilled professional technician who has the "know-how" required to successfully work on today's modern vehicles.

EXAM ADMINISTRATION

> *Note:* After November 2011, ASE will no longer offer paper and pencil certification exams. There will be no Winter testing window in 2012, and ASE will offer and support CBT testing exclusively starting in April 2012.

ASE provides computer-based testing (CBT) exams, which are administered at test centers across the nation. It is recommended that you go to the ASE website at *http://www.ase.com* and review the conditions and requirements for this type of exam. There is also an exam demonstration page that allows you to personally experience how this type of exam operates before you register.

CBT exams are available four times annually, for two-month windows, with a month of no testing in between each testing window:

- January/February – Winter testing window
- April/May – Spring testing window
- July/August – Summer testing window
- October/November – Fall testing window

Please note, testing windows and timing may change. It is recommended you go to the ASE website at *http://www.ase.com* and review the latest testing schedules.

UNDERSTANDING TEST QUESTION BASICS

ASE exam questions are written by service industry experts. Each question on an exam is created during an ASE-hosted "item-writing" workshop. During these workshops, expert service representatives from manufacturers (domestic and import), aftermarket parts and equipment manufacturers, working technicians, and technical educators gather to share ideas and convert them into actual exam questions. Each exam question written by these experts must then survive review by all members of the group. The questions are designed to address the practical application of repair and diagnosis knowledge and skills practiced by technicians in their day-to-day work.

After the item-writing workshop, all questions are pre-tested and quality-checked on a national sample of technicians. Those questions that meet ASE standards of quality and accuracy are included in the scored sections of the exams; the "rejects" are sent back to the drawing board or discarded altogether.

Depending on the topic of the certification exam, you will be asked between 40 and 80 multiple-choice questions. You can determine the approximate number of questions you can expect to be asked during the Brakes (A5) certification exam by reviewing the task list in Section 4 of this book. The five-year recertification exam will cover this same content; however, the number of questions for each content area of the recertification exam will be reduced by approximately one-half.

> *Note:* Exams may contain questions that are included for statistical research purposes only. Your answers to these questions will not affect your score, but since you do not know which ones they are, you should answer all questions in the exam.

Using multiple criteria, including cross-sections by age, race, and other background information, ASE is able to guarantee that exam questions do not include bias for or against any particular group. A question that shows bias toward any particular group is discarded.

TEST-TAKING STRATEGIES

Before beginning your exam, quickly look over the exam to determine the total number of questions that you will need to answer. Having this knowledge will help you manage your time throughout the exam to ensure you have enough available to answer all of the questions presented. Read through each question completely before marking your answer. Answer the questions in the order they appear on the exam. Leave the questions blank that you are not sure of and move on to the next question. You can return to those unanswered questions after you have finished the others. These questions may actually be easier to answer at a later time once your mind has had additional time to consider them on a subconscious level. In addition, you might find information in other questions that will help you recall the answers to some of them.

Multiple-choice exams are sometimes challenging because there are often several choices that may seem possible, or partially correct, and therefore it may be difficult to decide on the most appropriate answer choice. The best strategy, in this case, is to first determine the correct answer before looking at the answer options. If you see the answer you decided on, you should still be careful to examine the other answer options to make sure that none seems more correct than yours. If you do not know or are not sure of the answer, read each option very carefully and try to eliminate those options that you know are incorrect. That way, you can often arrive at the correct choice through a process of elimination.

If you have gone through the entire exam, and you still do not know the answer to some of the questions, *then guess.* Yes, guess. You then have at least a 25 percent chance of being correct. While your score is based on the number of questions answered correctly, any question left blank, or unanswered, is automatically scored as incorrect.

There is a lot of "folk" wisdom on the subject of test taking that you may hear about as you prepare for your ASE exam. For example, there are those who would advise you to avoid response options that use certain words such as *all, none, always, never, must,* and *only,* to name a few. This, they claim, is because nothing in life is exclusive. They would advise you to choose response options that use words that allow for some exception, such as *sometimes, frequently, rarely, often, usually, seldom,* and *normally.* They would also advise you to avoid the first and last option (A or D) because exam writers, they feel, are more comfortable if they put the correct answer in the middle (B or C) of the choices. Another recommendation often offered is to select the option that is either shorter or longer than the other three choices because it is more likely to be correct. Some would advise you to never change an answer since your first intuition is usually correct. Another area of "folk" wisdom focuses specifically on any repetitive patterns created by your question responses (e.g., A, B, C, A, B, C, A, B, C).

Many individuals may say that there are actual grains of truth in this "folk" wisdom, and whereas with some exams, this may prove true, it is not relevant in regard to the ASE certification exams. ASE validates all exam questions and test forms through a national sample of technicians, and only those questions and test forms that meet ASE standards of quality and accuracy are included in the scored sections of the exams. Any biased questions or patterns are discarded altogether, and therefore, it is highly unlikely you will experience any of this "folk" wisdom on an actual ASE exam.

PREPARING FOR THE EXAM

Delmar, Cengage Learning wants to make sure we are providing you with the most thorough preparation guide possible. To demonstrate this, we have included hundreds of preparation questions in this guide. These questions are designed to provide as many opportunities as possible to prepare you to successfully pass your ASE exam. The preparation approach we recommend and outline in this book is designed to help you build confidence in demonstrating what task area content you already know well while also outlining what areas you should review in more detail prior to the actual exam.

We recommend that your first step in the preparation process should be to thoroughly review Section 3 of this book. This section contains a description and explanation of the type of questions you'll find on an ASE exam.

Once you understand how the questions will be presented, we then recommend that you thoroughly review Section 4 of this book. This section contains information that will help you establish an understanding of what the exam will be evaluating, and specifically, how many questions to expect in each specific task area.

As your third preparatory step, we recommend you complete your first preparation exam, located in Section 5 of this book. Answer one question at a time. After you answer each question, review the answer and question explanation information located in Section 6. This section will provide you with instant response feedback, allowing you to gauge your progress, one question at a time, throughout this first preparation exam. If after reading the question explanation you do not feel you understand the reasoning for the correct answer, go back and review the task list overview (Section 4) for the task that is related to that question. Included with each question explanation is a clear identifier of the task area that is being assessed (e.g., Task A.1). If at that point you still do not feel you have a solid understanding of the material, identify a good source of information on the topic, such as an educational course, textbook, or other related source of topical learning, and do some additional studying.

After you have completed your first preparation exam and have reviewed your answers, you are ready to complete your next preparation exam. A total of six practice exams are available in Section 5 of this book. For your second preparation exam, we recommend that you answer the questions as if you were taking the actual exam. Do not use any reference material or allow any interruptions in order to get a feel for how you will do on the actual exam. Once you have answered all of the questions, grade your results using the Answer Key in Section 6. For every question that you gave an incorrect answer to, study the explanations to the answers and/or the overview of the related task areas. Try to determine the root cause for missing the question. The easiest thing to correct is learning the correct technical content. The hardest things to correct are behaviors that lead you to an incorrect conclusion. If you knew the information but still got the question incorrect, there is likely a test-taking behavior that will need to be corrected. An example of this would be reading too quickly and skipping over words that affect your reasoning. If you can identify what you did that caused you to answer the question incorrectly, you can eliminate that cause and improve your score.

Here are some basic guidelines to follow while preparing for the exam:

- Focus your studies on those areas you are weak in.
- Be honest with yourself when determining if you understand something.
- Study often but for short periods of time.
- Remove yourself from all distractions when studying.
- Keep in mind that the goal of studying is not just to pass the exam; the real goal is to learn.
- Prepare physically by getting a good night's rest before the exam, and eat meals that provide energy but do not cause discomfort.
- Arrive early to the exam site to avoid long waits as test candidates check in.
- Use all of the time available for your exams. If you finish early, spend the remaining time reviewing your answers.
- Do not leave any questions unanswered. If absolutely necessary, guess. All unanswered questions are automatically scored as incorrect.

Here are some items you will need to bring with you to the exam site:

- A valid government or school-issued photo ID
- Your test center admissions ticket
- A watch (not all test sites have clocks)

Note: Books, calculators, and other reference materials are not allowed in the exam room. The exceptions to this list are English-Foreign dictionaries or glossaries. All items will be inspected before and after testing.

WHAT TO EXPECT DURING THE EXAM

When taking a CBT exam, as soon as you are seated in the testing center, you will be given a brief tutorial to acquaint you with the computer-delivered test prior to taking your certification exam(s). The CBT exams allow you to select only one answer per question. You can also change your answers as many times as you like. When you select a second answer choice, the CBT will automatically unselect your first answer choice. If you want to skip a question to return to later, you can utilize the "flag" feature, which will allow you to quickly identify and review questions whenever you are ready. Prior to completing your exam, you will also be provided with an opportunity to review your answers and address any unanswered questions.

TESTING TIME

Each individual ASE CBT exam has a fixed time limit. Individual exam times will vary based upon exam area, and will range anywhere from a half hour to two hours. You will also be given an additional 30 minutes beyond what is allotted to complete your exams to ensure you have adequate time to perform all necessary check-in procedures, complete a brief CBT tutorial, and potentially complete a post-test survey.

You can register for and take multiple CBT exams during one testing appointment. The maximum time allotment for a CBT appointment is four and a half hours. If you happen to register for so many exams that you will require more time than this, your exams will be scheduled into multiple appointments. This could mean that you have testing on both the morning and afternoon of the same day, or they could be scheduled on different days, depending on your personal preference and the test center's schedule.

It is important to understand that if you arrive late for your CBT test appointment, you will not be able to make up any missed time. You will only have the scheduled amount of time remaining in your appointment to complete your exam(s).

Also, while most people finish their CBT exams within the time allowed, others might feel rushed or not be able to finish the test, due to the implied stress of a specific, individual time limit allotment. Before you register for the CBT exams, you should review the number of exam questions that will be asked along with the amount of time allotted for that exam to determine whether you feel comfortable with the designated time limitation or not.

As an overall time management recommendation, you should monitor your progress and set a time limit you will follow with regard to how much time you will spend on each individual exam question. This should be based on the total number of questions you will be answering.

Also, it is very important to note that if for any reason you wish to leave the testing room during an exam, you must first ask permission. If you happen to finish your exam(s) early and wish to leave the testing site before your designated session appointment is completed, you are permitted to do so only during specified dismissal periods.

UNDERSTANDING HOW YOUR EXAM IS SCORED

You can gain a better perspective about the ASE certification exams if you understand how they are scored. ASE exams are scored by an independent organization having no vested interest in ASE or in the automotive industry. With CBT exams, you will receive your exam scores immediately.

Each question carries the same weight as any other question. For example, if there are 50 questions, each is worth 2 percent of the total score.

The passing grade is 70 percent. That means you must correctly answer 35 out of the 50 questions to pass the exam. Your exam results can tell you

- Where your knowledge equals or exceeds that needed for competent performance, or
- Where you might need more preparation.

Your ASE exam score report is divided into content "task" areas; it will show the number of questions in each content area and how many of your answers were correct. These numbers provide information about your performance in each area of the exam. However, because there may be a different number of questions in each content area of the exam, a high percentage of correct answers in an area with few questions may not offset a low percentage in an area with many questions.

It should be noted that one does not "fail" an ASE exam. The technician who does not pass is simply told "More Preparation Needed." Though large differences in percentages may indicate problem areas, it is important to consider how many questions were asked in each area. Since each exam evaluates all phases of the work involved in a service specialty, you should be prepared in each area. A low score in one area could keep you from passing an entire exam. If you do not pass the exam, you may take it again at any time it is scheduled to be administered.

There is no such thing as average. You cannot determine your overall exam score by adding the percentages given for each task area and dividing by the number of areas. It does not work that way because there generally are not the same number of questions in each task area. A task area with 20 questions, for example, counts more toward your total score than a task area with 10 questions.

Your exam report should give you a good picture of your results and a better understanding of your strengths and areas needing improvement for each task area.

Types of Questions on an ASE Exam

SECTION 3

Understanding not only what content areas will be assessed during your exam, but how you can expect exam questions to be presented will enable you to gain the confidence you need to successfully pass an ASE certification exam. The following examples will help you recognize the types of question styles used in ASE exams and assist you in avoiding common errors when answering them.

Most initial certification tests are made up of between 40 and 80 multiple-choice questions. The five-year recertification exams will cover the same content as the initial exam; however, the actual number of questions for each content area will be reduced by approximately one-half. Refer to Section 4 of this book for specific details regarding the number of questions to expect during the initial Brakes (A5) certification exam.

Multiple-choice questions are an efficient way to test knowledge. To correctly answer them, you must consider each answer choice as a possibility, and then choose the answer choice that *best* addresses the question. To do this, read each word of the question carefully. Do not assume you know what the question is asking until you have finished reading the entire question.

About 10 percent of the questions on an actual ASE exam will reference an illustration. These drawings contain the information needed to correctly answer the question. The illustration should be studied carefully before attempting to answer the question. When the illustration is showing a system in detail, look over the system and try to figure out how the system works before you look at the question and the possible answers. This approach will ensure that you do not answer the question based upon false assumptions or partial data, but instead have reviewed the entire scenario being presented.

MULTIPLE-CHOICE/DIRECT QUESTIONS

The most common type of question used on an ASE exam is the direct multiple-choice style question. This type of question contains an introductory statement, called a stem, followed by four options: three incorrect answers, called distracters, and one correct answer, the key.

When the questions are written, the point is to make the distracters plausible to draw an inexperienced technician to inadvertently select one of them. This type of question gives a clear indication of the technician's knowledge.

Here is an example of a direct style question:

1. Which of the following would best describe the burnishing procedure?
    A. Four complete stops from 25 MPH.
    B. Twenty slow-downs from 50 MPH to 30 MPH.
    C. Ten slow-downs from 50 MPH to 30 MPH.
    D. Fifty slow-downs from 40 MPH to 15 MPH.

TASK C.15

Answer A is incorrect. This is not enough to heat the brakes thoroughly.

Answer B is correct. Twenty slow-downs from 50 MPH to 30 MPH would be an acceptable break-in method.

Answer C is incorrect. This is not enough to heat the brakes.

Answer D is incorrect. This would overheat the brakes.

COMPLETION QUESTIONS

A completion question is similar to the direct question except the statement may be completed by any one of the four options to form a complete sentence. Here's an example of a completion question:

TASK A.4.2

2. The brake system can be flushed with:
   A. Demineralized water.
   B. Mineral spirits.
   C. Brake fluid.
   D. Transmission fluid.

Answer A is incorrect. Water should not be introduced into the brake system. Brake fluid is hydroscopic; when brake fluid absorbs moisture the boiling temperature is lowered.

Answer B is incorrect. Mineral spirits contains petroleum distillates. When it comes into contact with brake fluid it will cause rubber components in the braking system to swell.

Answer C is correct. Using clean brake fluid will effectively flush the brake hydraulic system.

Answer D is incorrect. Transmission fluid contains oil which, when combined with brake fluid, will cause the rubber components to swell.

TECHNICIAN A, TECHNICIAN B QUESTIONS

This type of question is usually associated with an ASE exam. It is, in fact, two true-false statements grouped together, such as: "Technician A says ..." and "Technician B says ...", followed by "Who is correct?"

In this type of question, you must determine whether either, both, or neither of the statements are correct. To answer this type of question correctly, you must carefully read each technician's statement and judge it on its own merit.

Sometimes this type of question begins with a statement about some analysis or repair procedure. This statement provides the setup or background information required to understand the conditions about which Technician A and Technician B are talking, followed by two statements about the cause of the concern, proper inspection, identification, or repair choices.

Analyzing this type of question is a little easier than the other types because there are only two ideas to consider, although there are still four choices for an answer.

Again, Technician A, Technician B questions are really double true-or-false statements. The best way to analyze this type of question is to consider each technician's statement separately. Ask yourself, "Is A true or false? Is B true or false?" Once you have completed an individual evaluation of each statement, you will have successfully determined the correct answer choice for the question, "Who is correct?".

An important point to remember is that an ASE Technician A, Technician B question will never have Technician A and B directly disagreeing with each other. That is why you must evaluate each statement independently.

An example of a Technician A/Technician B style question looks like this:

3. The rear wheels on a non-ABS car equipped with front disc/rear drum brakes lockup when the vehicle is under hard braking. Technician A says this could be caused by front brakes which are not operating. Technician B says this could be caused by a malfunctioning metering valve. Who is correct?

    A. A only
    B. B only
    C. Both A and B
    D. Neither A nor B

TASK A.3.1

Answer A is incorrect. Technician B is also correct.

Answer B is incorrect. Technician A is also correct.

Answer C is correct. Both technicians are correct. If the front brakes are not operating then the rear brakes will be required to do much more work which will tend to cause them to lockup. The metering valve is supposed to hold off the front brakes until the rear brakes apply. If the valve did not let the front brakes apply, then the rear brakes would be overworked and tend to lockup.

Answer D is incorrect. Both technicians are correct

EXCEPT QUESTIONS

Another type of question used on ASE exams contains answer choices that are all correct except for one. To help easily identify this type of question, whenever it is presented in an exam, the word "EXCEPT" will always be displayed in capital letters. Furthermore, a cautionary statement will alert you to the fact that the next question is different from the ones otherwise found in the exam. With the EXCEPT type of question, only one *incorrect* choice will actually be listed among the options, and that incorrect choice will be the key to the question. That is, the incorrect statement is counted as the correct answer for that question.

Be careful to read these question types slowly and thoroughly; otherwise, you may overlook what the question is actually asking and answer the question by selecting the first correct statement.

An example of this type of question would appear as follows:

4. All of the following could cause the red brake warning light to come on EXCEPT:

    A. A corroded wheel cylinder.
    B. A leaking wheel cylinder.
    C. A leaking disc brake caliper.
    D. A bypassing master cylinder.

TASK A.3.4

Answer A is correct. A corroded wheel cylinder would prevent the brake from applying at that wheel; however, the red brake warning light is turned on by a pressure differential. A stuck wheel cylinder will not cause a pressure differential.

Answer B is incorrect. A leaking wheel cylinder will cause a pressure differential.

Answer C is incorrect. A leaking disc brake caliper will cause a pressure differential

Answer D is incorrect. A bypassing master cylinder will cause a pressure differential.

LEAST LIKELY QUESTIONS

LEAST LIKELY questions are similar to EXCEPT questions. Look for the answer choice that would be the LEAST LIKELY cause (most incorrect) for the described situation. To help easily identify these types of questions, whenever they are presented in an exam the words "LEAST LIKELY" will always be displayed in capital letters. In addition, you will be alerted before a LEAST LIKELY question is posed. Read the entire question carefully before choosing your answer.

An example of this type of question is shown here:

TASK A.4.1

5. Which of the following is LEAST LIKELY to cause an external brake fluid leak?
   A. A rusted line.
   B. A bypassing master cylinder.
   C. Poor maintenance habits.
   D. A ruptured flexible hose.

Answer A is incorrect. A rusty line could develop a hole and cause a brake fluid leak.

Answer B is correct. A bypassing master cylinder will allow fluid to go back up into the master cylinder but would not cause an external leak.

Answer C is incorrect. Poor maintenance habits such as failing to flush the brake system at required intervals can lead to internal rust and brake fluid leaks.

Answer D is incorrect. A ruptured flexible hose will allow an external brake fluid leak.

SUMMARY

The question styles outlined in this section are the only ones you will encounter on any ASE certification exam. ASE does not use any other types of question styles, such as fill-in-the-blank, true/false, word-matching, or essay. ASE also will not require you to draw diagrams or sketches to support any of your answer selections, although any of the described question styles may include illustrations, charts, or schematics to clarify a question. If a formula or chart is required to answer a question, it will be provided for you. You may in rare cases be required to solve a simple math problem, so bringing a simple pocket calculator to the test session might be a good idea.

SECTION 4

Task List Overview

INTRODUCTION

This section of the book outlines the content areas or *task list* for this specific certification exam, along with a written overview of the content covered in the exam.

The task list describes the actual knowledge and skills necessary for a technician to successfully perform the work associated with each skill area. This task list is the fundamental guideline you should use to understand what areas you can to expect to be tested on, as well as how each individual area is weighted to include the approximate number of questions you can expect to be given for that area during the ASE certification exam. It is important to note that the number of exam questions for a particular area is to be used as a guideline only. ASE advises that the questions on the exam may not equal the number listed on the task list. The task lists are specifically designed to tell you what ASE expects you to know how to do and to help prepare you to be tested.

Similar to the role this task list will play in regard to the actual ASE exam, Delmar, Cengage Learning has developed six preparation exams, located in Section 5 of this book, using this task list as a guide. It is important to note that although both ASE and Delmar, Cengage Learning use the same task list as a guideline for creating these test questions, none of the test questions you will see in this book will be found in the actual, live ASE exams. This is true for any test preparatory material you use. Real exam questions are *only* visible during the actual ASE exams.

Task List at a Glance

The Brakes (A5) task list focuses on six core areas, and you can expect to be asked a total of approximately 45 questions on your certification exam, broken out as outlined:

- A. Hydraulic System Diagnosis and Repair (12 questions)
    1. Master Cylinder (3)
    2. Lines and Hoses (3)
    3. Valves and Switches (3)
    4. Bleeding, Flushing, and Leak Testing (3)
- B. Drum Brake Diagnosis and Repair (5 questions)
- C. Disc Brake Diagnosis and Repair (10 questions)
- D. Power Assist Units Diagnosis and Repair (4 questions)
- E. Miscellaneous Systems Diagnosis and Repair (7 questions)
- F. Electronic Brake Control Systems – Antilock Brake Systems (ABS), Traction Control Systems (TCS), and Electronic Stability Control Systems (ESC) – Diagnosis and Repair (7 questions)

Based upon this information, the following is a general guideline demonstrating which areas will have the most focus on the actual certification exam. This data may help you prioritize your time when preparing for the exam.

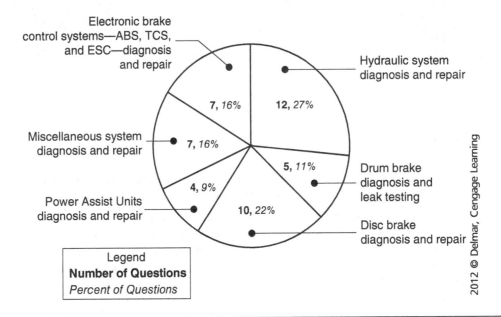

Note: The actual number of questions you will be given on the ASE certification exam may vary slightly from the information provided in the task list, as exams may contain questions that are included for statistical research purposes only. Do not forget that your answers to these research questions will not affect your score.

A. Hydraulic System Diagnosis and Repair (12 Questions)

1. Master Cylinder (3 Questions)

1. Diagnose poor stopping, dragging, high or low pedal, hard or spongy pedal caused by the master cylinder; determine needed repairs.

Modern brake systems consist of front disc brakes, rear drum brakes, master cylinder, lines, valves, power assist booster, brake pedal, and flexible hoses. Many vehicles today are equipped with front disc and rear disc brakes. For safe and effective brake operation, all components in the brake system must be in proper working order.

The heart of the brake system is the master cylinder. The master cylinder is a hydraulic pump that is operated by a pushrod connected to the brake pedal or assisted by a brake booster. The purpose of the master cylinder is to convert the driver's mechanical force into hydraulic pressure. Master cylinders are divided into two separate hydraulic systems. On most vehicles, one system controls the rear brakes and the other system controls the front brakes. If a hydraulic failure occurs in one system of a dual system master cylinder, the other system remains operational. This provides greater protection against a total hydraulic brake failure.

Some master cylinders are split diagonally. In this system one half of the master cylinder controls one front wheel on one side and one rear wheel on the other side. The diagonal split system is most often found on front-wheel drive vehicles. Master cylinders are manufactured aluminum and have a separate plastic reservoir. Aluminum master cylinders are anodized, which is intended to extend the life of the master cylinder by reducing corrosion.

There are several common symptoms of master cylinder related problems. An improperly adjusted brake light switch or cruise control switch will cause brake drag by keeping the brake pedal partially depressed. This causes the master cylinder to maintain pressure on the brake system. Also, a binding brake pedal caused by worn pivot bushings can cause brake drag.

Air in the hydraulic system will cause a low, spongy pedal. An internal leak in the master cylinder cups may also cause a low, sinking pedal. Always consider the level and condition of the brake fluid when diagnosing master cylinder problems. Refer to Task A.4.1 and A.4.4 for more on brake fluid. A worn or corroded internal master cylinder bore may cause excessive pedal effort and brake drag. Another cause for brake drag is a blocked master cylinder compensation port or vent port.

Brake drag will cause overheated brake linings. It is important to inspect the entire brake system after determining a master cylinder problem.

2. Diagnose problems in the stepbore/quick take-up master cylinder and internal valves (e.g., volume control devices, quick take-up valve, fast-fill valve, pressure regulating valve); determine needed repairs.

The step bore, or quick take-up, master cylinder incorporates two different size bore diameters. The step bore master cylinder improves brake effectiveness in low drag disc brake calipers, which improves fuel economy. In a step bore master cylinder, the forward bore is slightly smaller than the rear bore. This design allows the small bore of the cylinder to supply high pressure with low volume. The large bore supplies high volume at low pressure. The large bore system initially supplies a large volume of fluid to move the caliper piston and apply the brake pads. Once the pads are in contact with the brake rotors the small bore applies higher hydraulic pressure.

The aluminum bore of a step bore master cylinder is anodized and cannot be honed. Honing would remove the hard bore coating. If the bore is pitted or damaged, the master cylinder must be replaced.

Step bore master cylinders must be bench bled due to severe mounting angle, the quick take-up valve, and the four-port step bore design. It is nearly impossible to bleed all the air from the master cylinder once it is installed.

Diagnosing problems with a step bore master cylinder requires attention to details that are not common to other types of master cylinders. Never overfill a step bore master cylinder. Overfilling can cause the front brakes to drag or cause excessive brake pedal effort. Another cause for excessive pedal effort or a hard pedal on a step bore master cylinder is a clogged or plugged quick take-up valve. On the other hand, if a quick take-up valve opens too early, this can cause a low pedal condition.

A defective quick take-up valve could cause a front caliper to drag and result in premature, uneven pad wear.

3. Measure and adjust master cylinder pushrod length.

Brake pedal height and pushrod adjustment are often overlooked when diagnosing a brake problem. Correct pushrod adjustment is very important for proper braking. The pushrod must have a very slight gap between the tip of the pushrod and its mating surface in the master cylinder. Most manufacturers' pushrod length can be adjusted by unlocking and rotating an adjustment nut that is on the tip of the pushrod. A pushrod that is too long may cause the master cylinder piston to over-travel. This may damage the pistons seals and lead to a master cylinder failure. A pushrod that is too long may also cause the piston seals to block off the compensating (vent) port, which may cause brake drag or lockup. If a vehicle has had repeated master cylinder failures, verify that the pushrod is not over-extended. A pushrod that is too short may limit the amount of brake pressure developed and reduce the overall braking ability of the vehicle.

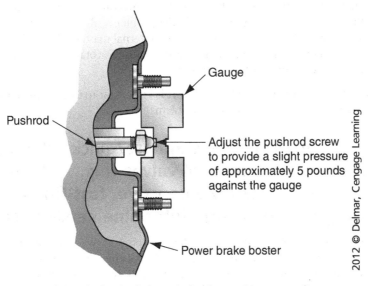

Always check the brake light switch after making any adjustments.

4. Check master cylinder for failures by depressing brake pedal; determine needed repairs.

Three common complaints that can be caused by a faulty master cylinder are brake fade, spongy pedal, and excessive pedal effort. Before condemning the master cylinder, a complete brake inspection of all components must be performed.

If the brake pedal slowly sinks to the floor when applied and the vehicle is stopped with the engine running, there may be a leak in master cylinder cups. This is known as "bypassing." Air leaking past the master cylinder piston cups will cause air to enter the fluid and cause a spongy pedal.

Swollen master cylinder cups may cause excessive pedal effort or can be the result of a corroded master cylinder bore. If the master cylinder reservoir cover seal is badly swollen, suspect brake fluid contamination from oil, transmission fluid, power steering fluid, etc. If fluid contamination is discovered, a complete inspection of the brake system must be performed. Contaminated fluid will damage all rubber components and seals in the brake calipers, wheel cylinders, and other brake parts. The technician must flush all the contaminated fluid from the brake system and replace all seals and cups in the master cylinder, calipers, and wheel cylinders. In addition, any time the lines are disconnected from the master cylinder, or any other component, the brake system must be bled to remove trapped air from the lines and master cylinder.

5. Diagnose the cause of master cylinder external fluid leakage.

External leaks from the master cylinder are usually easy to diagnose. Thoroughly clean the master cylinder first and check the fluid level. Check the condition of the reservoir seal and make sure the cover snaps tightly into place. If fluid level is low it may be normal due to brake pad wear or it may be an indication of a leak.

A complete visual inspection of the entire brake system must be performed to accurately determine any problems. If there is excessive fluid at the rear of the master cylinder, or if the paint on the booster is blistered, check for leaking brake fluid at the rear of the master cylinder. If a leak is present where the booster bolts up to the master cylinder, there will probably be brake fluid in the booster. The booster must be inspected and cleaned thoroughly or will have to be replaced.

Some master cylinders have a brake fluid reservoir, which is fitted into the main body of the cylinder. The seals of the reservoir and the cap must be inspected for possible leaks.

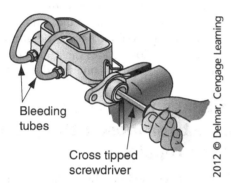

6. Remove and replace master cylinder; bench bleed and test operation and install master cylinder; verify master cylinder function.

Replacing the master cylinder is usually straightforward. Start by cleaning the exterior of the master cylinder and the area around the master cylinder. Remove the brake lines and plug the ports on the master cylinder and the brake lines. Disconnect any electrical wiring to the master cylinder. Remove the attaching nuts at the brake booster. If the vehicle has manual brakes, disconnect the pushrod from the brake pedal. Carefully lift the old master cylinder away from the vehicle to avoid spilling brake fluid on the vehicle's exterior. Use fender covers to avoid brake fluid damage to the paint. If brake fluid comes in contact with paint, wash immediately with water.

Before the replacement master cylinder is installed, it must be bench bled. Firmly secure the master cylinder in a vise and fill the reservoir with clean brake fluid. Make sure the master cylinder is level. Usually the new or rebuilt master will come with a pair of plastic fittings and hoses. Screw the fittings into the master cylinder ports. Attach one end of the hose to each port and place the other end of the hose into the reservoir. Make sure that both hoses remain submerged in the brake fluid of the reservoir. Using a drift or dowel, slowly press the master cylinder piston in and out until the air is dispelled from the cylinder.

Step bore master cylinders require 75 to 100 psi to open the quick take-up valve. An option is to use a bleeder syringe to suck the air from the rear step bore of the master.

Mount the master cylinder back into the vehicle. Check pedal height and brake pedal free play. The entire brake system must be bled after the master is installed to insure all the air is purged from the lines. Road test the vehicle and verify proper master cylinder and brake operation.

2. Lines and Hoses (3 Questions)

1. Diagnose poor stopping, pulling, or dragging caused by problems in the lines and hoses; determine needed repairs.

The master cylinder is connected to the wheel cylinders and calipers by a network of brake lines and hoses. Brake hoses create a flexible link between the steel line at the frame or body and the brake calipers or wheel cylinders.

The pressure and flow throughout the entire brake system must be within specification. Even a partial blockage in the lines or hoses can restrict the flow of fluid and cause problems. The applying brake pressure may be enough to force fluid through a restriction, but the affected brake may not release or will release slowly. This can cause brake drag. When the restriction affects one wheel, this can cause a pull to one side during braking.

2. Inspect brake lines and fittings for leaks, dents, kinks, rust, cracks, or wear; inspect for loose fittings and supports; determine needed repairs.

A comprehensive brake inspection of the brake lines and brake hoses should be part of any brake system diagnosis.

Kinks or dents in the brake lines may affect hydraulic pressure. Also inspect all fittings for leaks and secure connections. Check all attaching supports at the frame, undercarriage, rear suspension, and struts. Attaching brackets and supports may loosen up or corrode over time. This may cause a shift in a brake line or hose and may cause damage to a line. A major concern for older vehicles, and in areas where salt is used during winter months, is corrosion and rust. Rusted brake lines can actually become porous and eventually leak brake fluid. Should this occur, a new line will need to be installed, or the damaged section replaced. When repairing sections of brake line, the parts of the old line that were cut will need to be double flared and connected to the new section of line using unions. Never use compression fittings to connect two pieces of brake line.

3. Inspect flexible brake hoses for leaks, kinks, cracks, bulging, wear, or corrosion; inspect for loose fittings and supports; determine needed repairs.

Because flexible brake hoses move with the natural movement of the steering and suspension they must be inspected for wear. Front flexible brake hoses are especially susceptible to cracking and wear over time, as the wheels are turned left and right. Brake hoses should be inspected for bulging and wear marks caused by abrasion.

Check for the improper installation of brake hoses. Brake hoses twisted at calipers or loose brake hoses at the attaching supports are problems you may encounter. Whenever a brake hose is replaced make sure a new sealing washer is used.

4. Replace brake lines, hoses, fittings, and supports; fabricate brake lines using proper material and flaring procedures (double flare and ISO types).

The ends of brake lines are either a double flared design or have International Organization for Standardization (ISO) type flared fittings. Copper tubing should never be used in a brake system. Copper is subject to fatigue and corrosion over time and can result in a hydraulic failure.

Factory replacement brake lines are often available. These lines are an exact fit with the proper end fittings. When brake lines are not available it will be necessary to fabricate a new line. Replacement lines come in various lengths and diameter. Carefully choose the correct size brake line with the correct end fittings. Remove the old line and use it as template to fabricate the new line. A bending tool will be needed to bend tight angles. Although brake lines bend easily, be careful not to form any kinks in the process. Remember, always make sure of the end fittings: either double flared or ISO-type flare. Also, watch for use of metric and/or fractional fittings on double flared installations.

When it becomes necessary to flare the ends of the brake line, one of two tools will be needed: the double flaring tool or the ISO flare-forming tool. Never use compression fittings to join sections of brake line.

Always use a flare nut wrench to loosen or tighten brake line fittings. An open-end wrench can damage the end by rounding the corners of the fitting nut. Make sure all supports are intact. Loose brake lines or hoses may result in damage or failure.

The entire brake system must be bled after installing a new hose or line. Also apply pressure to the brakes and inspect the repair for leaks. Road test the vehicle and confirm proper brake operation.

5. Inspect brake lines and hoses for proper routing and support.

All brake lines and hoses must be routed correctly. Inspect all lines and hoses and make certain that they are routed away from components that may interfere or rub, which can cause damage. Also, make sure brake lines and hoses are not routed near any components that generate excessive heat, such as an exhaust manifold, exhaust pipe, or catalytic converter.

Brake lines and hoses must be securely fastened to their supports. Unnecessary movement or vibration can lead to fatigue and eventual failure.

When replacing a flexible brake hose, make sure that it meets original equipment manufacturer (OEM) safety standards and that the hose is the exact size as the original. A hose that is too long may rub on a chassis or steering component. If the hose is too short, it may stretch and break when the moveable component reaches the end of its travel. Always replace the sealing washer every time a brake hose is serviced.

3. Valves and Switches (3 Questions)

1. Diagnose poor stopping, pulling, or dragging caused by problems in the hydraulic system valve(s); determine needed repairs.

To provide evenly balanced braking and to warn of impending problems, many different valves are used in modern hydraulic brake systems. The most common valves are the metering valve, the proportioning valve, and the pressure differential valve. These valves are sometimes housed in one assembly called the combination valve.

Faulty valves can cause brake drag, rear wheel lockup, or front wheel lockup upon braking. An in-depth review of these valves will follow in the upcoming paragraphs.

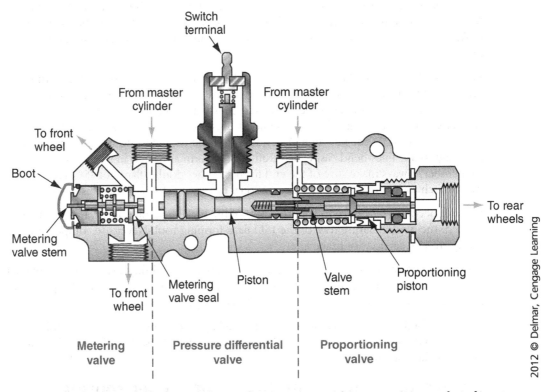

2. Inspect, test, and replace metering, proportioning, pressure differential, and combination valves.

The metering valve is used on vehicles with a front disc and rear drum brake design. The metering valve delays the front disc brake operation until the rear drum brake shoes overcome the return spring tension. This will ensure simultaneous brake application of the front and rear brakes. Without this valve the front disc brakes would apply too quickly.

A faulty metering valve can cause nose-diving and front wheel lockup when the brakes are applied. During this condition, too much initial brake pressure is being applied to the front disc brakes. Premature front disc pad wear will also result from a faulty metering valve. Inspect the valve for leaks.

The proportioning valve is also found on brake systems with front discs and rear drums. The proportioning valve controls and limits the pressure to the rear brakes. Under mild braking, the pressure to all wheels is about the same. As pedal effort increases, the possibility of rear wheel lockup exists. The proportioning valve prevents rear wheel lockup under heavy braking. Vehicles with split diagonal brake systems utilize dual proportioning valves.

A malfunctioning proportioning valve may cause rear wheel lockup under heavy braking, which may result in loss of vehicle control.

The pressure differential valve is used to warn the driver of a hydraulic pressure failure. The valve also incorporates a switch that will close and illuminate the dash brake warning light. A small piston floats inside the pressure differential valve cylinder, which separates the two halves of the brake system. The piston is normally centered inside the valve. When one side of the brake system develops a serious leak, the piston will be forced to the low-pressure side.

After the vehicle is repaired and brakes bled, the pressure differential valve should reset automatically. If the dash warning light does not go out, the valve may have to be re-centered. Check manufacturer's recommended procedure for centering the valve.

A combination valve incorporates the metering valve, proportioning valve, and the pressure differential valve all in one assembly. If any function of the combination valve fails, the entire assembly must be replaced.

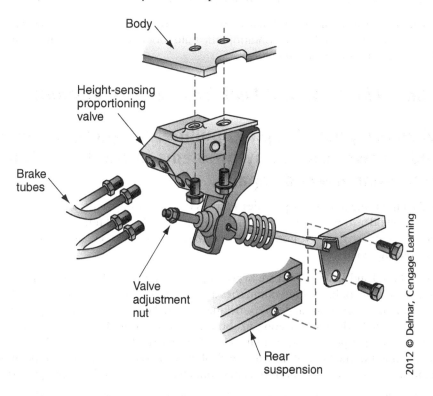

3. Inspect, test, replace, and adjust load or height sensing-type proportioning valve(s).

The load-sensing proportioning valve is mounted on the chassis, and a linkage is connected from the valve to the rear axle. When the load is light, the linkage positions the internal valve so brake pressure is reduced to the rear wheels during moderate brake applications. A heavy load on the rear suspension reduces rear chassis height. This action causes the linkage to move the valve in the height-sensing proportioning valve. Under this condition, the proportioning valve does not reduce pressure to the rear brakes during moderate brake application.

One type of load-sensing proportioning valve uses a bracket attached to the axle. A spring connects the bracket to a lever on the end of the proportioning valve, which is mounted to the vehicle chassis.

Weight transfer during hard braking will cause a change in chassis height. The increase in chassis height will decrease pressure to the rear wheels.

Any modification that alters the ride height will affect how the height-sensing-type proportioning valve operates.

4. Inspect, test, and replace brake warning light, switch, sensor, and circuit.

When the pressure is equal in the primary and secondary sections of the master cylinder, the warning switch piston remains centered. In this position, the switch piston does not touch the switch pin. If the pressure is unequal between the primary and secondary master cylinder sections, the pressure difference moves the switch piston to one side. In this position, the switch piston pushes the spring-loaded switch pin upward and closes the warning light switch. This action illuminates the brake warning light.

The brake warning lamp circuit may be tested by grounding the warning switch with the ignition switch on. Under this condition, the bulb should light. If the bulb is not lit, check the fuse, bulb, and connecting wires.

4. Bleeding, Flushing, and Leak Testing (3 Questions)

1. Diagnose poor stopping, pulling, dragging, or incorrect pedal travel caused by problems in the brake fluid; determine needed repairs.

Brake fluid is the liquid that provides a means of transmitting hydraulic pressure from the master cylinder to the calipers and wheel cylinders. Brake fluid must meet federal requirements for viscosity, non-corrosive quality, water tolerance, high boiling point, and low freezing point.

DOT 3 fluid and DOT 4 fluid are hygroscopic, which means they absorb moisture. Over time the quality of brake fluid diminishes and may lead to internal corrosion of brake components. Corrosion can lead to sticking or seized calipers and wheel cylinders, which may cause brake drag. Corrosion can also cause a wheel cylinder piston, master cylinder piston, or caliper piston to seize in its bore. The absorbed moisture can also lower the boiling point of the brake fluid which will allow it to evaporate at high temperatures and create air in the lines. The air in the lines could lead to a soft and/or spongy pedal.

Engine oil, transmission oil, or any fluid other than brake fluid will contaminate the brake system and may swell rubber components of the brake system and cause brake drag. If a brake system has been contaminated with a foreign substance it will be necessary to flush all the contamination from the brake system. Also, a careful inspection of the entire brake system is necessary because of the increased probability of damaged brake components.

2. Bleed and/or flush hydraulic system using manual, pressure, vacuum, or surge gravity method(s).

Brake systems function on the principle of hydraulics, which states that a liquid cannot be compressed. However, air is compressible and if air is present in the brake system, the result will be a low and spongy pedal. Also, air contains moisture which may lead to component corrosion.

Air will enter the brake system when a hydraulic component is disconnected or if the level of master cylinder becomes too low. Bleeding the brakes is necessary to remove all the air in the brake system. Never reuse brake fluid.

There are four common methods for bleeding brakes: manual bleeding, pressure bleeding, vacuum bleeding, and surge bleeding. Whatever method is chosen, some basic procedures are the same. Check the level of the fluid in the master cylinder and maintain that level throughout the bleeding process. On all vehicles, follow the manufacturer's bleeding sequence. Bleeder screws are provided at the master cylinder, wheel cylinders, disc brake calipers, and sometimes on the combination valve. Using these bleeder screws, air is purged from the system by opening these screws and forcing fluid through them.

A pressure bleeder has an adapter connected to the top of the master cylinder reservoir. A hose is connected from this adapter to the pressure bleeder chamber in the top of the pressure bleeder. The pressure bleeder has an air chamber below the fluid chamber, and a diaphragm that separates the air and fluid chambers. Shop air is used to pressurize the air chamber up to 15 to 20 psi (103 to 138 kPa). If the brake system has a metering valve, this valve may have to be held open with a special tool. The bleeder hose is connected from each bleeder screw into a container partially filled with brake fluid. Open the bleeder screw until a clear stream of brake fluid is discharged.

Some newer pressure bleeders have an electric motor/pump in them to create pressure, thus eliminating the need for shop air to be applied.

During the vacuum bleeding procedure, a vacuum pump is connected to a sealed container. The pump can be either handheld or powered by compressed shop air. With a handheld pump, another hose is connected from the container to the bleeder screw. A one-way check valve is connected in the hose from the bleeder screw to the container. Operate the pump handle 10 to 15 times to create vacuum in the container. Open the bleeder screw until about 1 inch (25 mm) of brake fluid is pulled into the container. Repeat the procedure until the fluid coming into the container is free of air bubbles. With the air powered vacuum bleeder, connect the hose from the device to the bleeder screw and open the bleeder. The vacuum created by the air powered bleeder will draw the air and brake fluid into the container. Continue to draw the fluid from the bleeder until no air can be seen. Be sure to keep the master cylinder fluid reservoir full of fluid during the entire procedure.

During a manual brake bleeding procedure, connect a bleeder hose from a bleeder screw into a container partially filled with brake fluid. Keep the end of this hose submerged below the level of the brake fluid in the container. Each wheel caliper or cylinder must be bled in the vehicle manufacturer's specified sequence in any bleeding procedure. Wheel calipers may be tapped with a soft hammer to help remove air bubbles. With the bleeder opened, have an assistant slowly press the brake pedal. Do not allow the master cylinder to bottom out as dirt or corrosion that may exist in the end of the master cylinder may damage the piston seals. With the pedal pressed, close the bleeder screw. The pedal can then be released. Repeat this process until no air comes out of the bleeder.

The surge method is a procedure that is sometimes used with manual brake bleeding. Surge bleeding is used in cases where it is difficult to remove trapped air from a wheel cylinder or brake caliper. In this method the pedal is first pumped several times. The bleeder screw is then opened and with a quick movement an assistant depresses the pedal. The pedal is then released slowly. Wait a few seconds and repeat. On the last down stroke of the pedal, close the bleeder quickly.

Flushing the brake system involves continuing the bleeding process until all the fluid in the brake system is completely replaced. Flushing is recommended when brake fluid has been contaminated and is also recommended as a periodic brake service. Brake fluid absorbs moisture over time. Flushing removes the moisture and maintains the integrity of the brake components. If major brake components are replaced, flushing is highly recommended.

3. Pressure test brake hydraulic system.

Using pressure gauges, the brake system can be checked for correct hydraulic pressure at different components. Following manufacturer's specification, it is possible to determine if the caliper, the wheel cylinder, or part of the brake system is not receiving the correct pressure. Using pressure gauges can help in accurately diagnosing failed components.

The proportioning valves connected to the rear wheels may be tested by connecting pressure gauges on the master cylinder side and wheel cylinder side of each proportioning valve. When the brake pedal is applied with light pressure and the master cylinder pressure is below the split point pressure, both gauges should indicate the same pressure. If the brake pedal pressure is increased and the master cylinder pressure exceeds the split point pressure, the master cylinder pressure should exceed the wheel cylinder pressure by the manufacturer's specified amount.

After the brakes are bled, fill the master cylinder to the proper level. Following several firm applications of the pedal, the calipers should be checked for leaks at the piston seals and the hose attachment areas.

4. Select, handle, store, and install proper brake fluids (including silicone fluids).

There are currently three types of brake fluid: DOT 3, DOT 4, and DOT 5. DOT 3 and DOT 4 are hygroscopic, which means they absorb moisture. All brake fluids must meet standards set by the Department of Transportation (DOT) and the Society of Automotive Engineers (SAE). The main difference of brake fluids is the point at which they begin to boil. DOT 3 is a glycol-based fluid that has a minimum boiling point of 401°F. DOT 4, which is also glycol-based, has a minimum boiling point of 446°F. DOT 5 brake fluid is silicone-based, is not hygroscopic, and has a minimum boiling point of 500°F.

Due to their hygroscopic nature, DOT 3 and DOT 4 fluids have a limited shelf life. When a container of brake fluid is opened, the contents must be used as soon as possible. Always replace the cap on the container immediately. DOT 3 and DOT 4 will damage a painted surface, so care must be taken when handling brake fluid.

The main advantage to DOT 5 silicon brake fluid is its high boiling point and the fact that it does not attract water. However, DOT 5 brake fluid is prone to aeration. Tiny air bubbles can form when the fluid is agitated, and for this reason is not typically used in ABS systems. DOT 5 fluid is also more difficult to bleed.

When adding brake fluid always use fresh fluid from a new container and always check for the correct fluid for that vehicle. Most vehicles use either DOT 3 or DOT 4 fluid. These fluids can be interchanged but should not be mixed. DOT 5 fluid is not compatible with DOT 3 or DOT 4.

B. Drum Brake Diagnosis and Repair (5 Questions)

1. Diagnose poor stopping, pulling, dragging, or incorrect pedal travel caused by drum brake hydraulic problems; determine needed repairs.

A springy and spongy pedal is a condition where the brake pedal does not give firm resistance to foot pressure and feels elastic. It is normally caused by air in the hydraulic lines. Poor stopping ability may be caused by hydraulic problems such as contaminated fluid or air in the hydraulic system.

Brake drag may be caused by hydraulic problems such as contaminated brake fluid, inferior rubber cups in the master cylinder or wheel cylinders, or plugged compensating (vent) ports in the master cylinder. An obstructed brake line or hose may also be the cause of brake drag. A restricted brake hose or line may also cause a pull to one side or reduce stopping ability.

A seized wheel cylinder piston can cause brake drag if the piston is stuck in its bore. If the piston is stuck outward it can cause brake drag and premature brake shoe wear. A leaking wheel cylinder can cause air to enter the system and loss of fluid. This will result in a low and spongy pedal. Poor quality brake fluid or overheated brakes may lead to fluid boiling. At this point the fluid can be compressed and a temporary loss of proper braking will occur.

2. Diagnose poor stopping, noise, pulling, grabbing, dragging, pedal pulsation, or incorrect pedal travel caused by drum brake mechanical problems; determine needed repairs.

Brake squeal may be caused by bent backing plates, distorted drums, loose linings at shoe ends, improper lining position on the shoe, weak or broken hold-down springs, excessive accumulations of brake dust, or loose wheel bearings. Brake chatter may be caused by improper brake adjustment; loose backing plates; contaminated linings; out-of-round, tapered, or barrel-shaped drums; cocked or distorted shoes; or loose wheel bearings.

Pedal pulsation may be caused by out-of-round brake drums.

Dragging or brake grabbing may be the result of contaminated brake shoes. Inspect the surface of the brake shoes for signs of brake fluid contamination or rear differential grease. Also, inspect wheel cylinders for seized pistons. If the wheel cylinder pistons do not retract when the brakes are not applied, this will result in brake drag. This drag will cause premature brake wear and overheated linings.

Incorrect pedal travel caused by drum brakes can usually be blamed on shoes that are out of adjustment. The shoes must travel further to contact the drums, which means that the driver must press the pedal further for the shoes to make contact.

3. Remove, clean, inspect, and measure brake drums; follow manufacturers' recommendations in determining need to machine or replace.

Brake drums should be inspected for cracks, heat checks, out-of-round, bell mouth, scoring, and hard spots. The inside drum diameter should be measured with a drum micrometer. If the drum diameter exceeds the maximum limit specified by the manufacturer, replace the drum. The drum diameter should be measured every 45 degrees around the drum. The maximum allowable out-of-round specified by some manufacturers is 0.0035 inch (0.09 mm). If the out-of-round exceeds specifications, the drum may be machined if the machining does not cause the drum to exceed the maximum allowable diameter.

4. Machine drums according to manufacturers' procedures and specifications.

When machining a brake drum, the drum must be installed securely and centered on the lathe. A dampening belt must be installed tightly around the outside of the drum to prevent the cutting tool from chattering on the drum. Many manufacturers recommend a rough cut tool of 0.005 to 0.010 inch (0.127 to 0.254 mm) and a finish cut tool depth of 0.005 inch (0.127 mm). After the machining procedure, the drum should be sanded to remove minor irregularities.

The surface of a freshly refinished drum contains millions of tiny particles. These particles can remain free on the surface or can become lodged in open pores of the cast iron drum. If these particles are not removed they can become imbedded in the linings. Once imbedded in the linings, the linings become a fine grinding stone, which will score the brake drum.

Drums must be washed thoroughly after machining. Soapy water works best. Simply submerge the drum to remove left over debris from machining. Brake clean and rags are not the preferred method because they usually leave behind metal debris that embeds in the new pads and can cause brake squeaks.

5. Using proper safety procedures, remove, clean, and inspect brake shoes/linings, springs, pins, self-adjusters, levers, clips, brake backing (support) plates, and other related brake hardware; determine needed repairs.

Before removing brake shoes, be sure to clean the assembly with an approved cleaning method that will contain all brake dust and prevent it from becoming airborne. Remove the brake shoes using the proper tools to prevent any personal injury that may result from improper methods. All brake return springs should be inspected for distortion and stretching. Brake shoes should be inspected for broken welds, cracks, wear, and distortion. If the wear pattern on the brake shoes is uneven, the shoes are distorted. Check all clips and levers for wear and bending. Inspect the brake linings for contamination with oil, grease, or brake fluid. If contamination is found, locate the source and correct the failure prior to new shoe installation. Clean and lubricate adjusting and self-adjusting mechanisms.

A bent backing plate will not align the brake shoes properly. The misaligned shoes will result in brake drag or grab. Loose backing plate anchor bolts may cause brake chatter because the brake shoes can become misaligned if the backing plate moves.

If the brake shoes and linings have a slight blue coloring, this indicates overheating. In this case, the brake shoe adjusting screw springs and brake shoe hold-down springs should be replaced.

Linings must be replaced when contaminated with grease, oil, or brake fluid. Operating a vehicle for any length of time with worn brake shoes and linings will quickly result in scored brake drums. Always replace parts that are near the end of their service life. Never replace linings on one brake assembly without replacing those on the opposite wheel.

To check for leaks, pull back each wheel cylinder dust boot. Be careful as some manufacturers do not recommend pulling back the wheel cylinder dust boots as this may allow debris inside and damage the piston seals. Normally you will see a small amount of brake fluid present. This is not a cause for alarm as it acts as a lubricant for the piston. However, large amounts of fluid behind the boot indicates the fluid is leaking past the piston cups and an overhaul or replacement is in order. Check the pads where the brake shoes rest on the backing plate. Look for any deep grooves in the brake shoe and lining contact pads that could resist brake shoe movement. Hand sand any grooves in the braking surface. If grooves are still present after sanding, the backing plate must be replaced. Any attempt to remove the grooves by grinding may result in improper brake shoe to brake drum contact.

With manual brake adjusters, the brakes must be periodically adjusted to limit pedal movement and the increased need for brake fluid. Because the wear is so gradual, the driver may not realize how much out of adjustment the brakes actually are.

Automatic brake adjusters generally take the worry out of keeping the brakes properly adjusted. The automatic adjusters automatically adjust the clearance between the shoe linings and the drum.

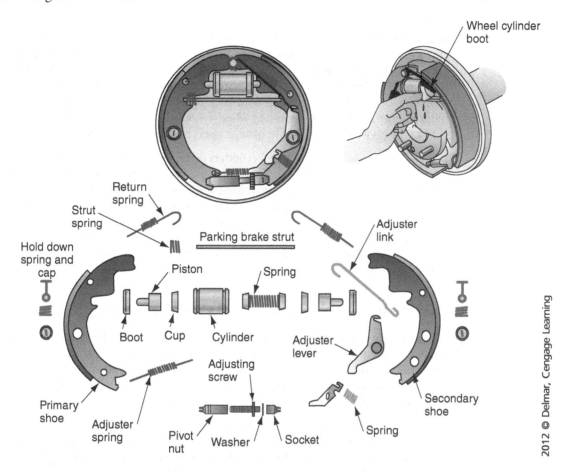

6. Lubricate brake shoe support pads on backing (support) plate, self-adjuster mechanisms, and other brake hardware.

Dry brake shoe support pads on the backing plate may cause a squeaking noise during brake applications. Lightly scored brake shoe support pads may be resurfaced and lubricated with high temperature grease.

Whenever the rear brake shoe and linings are removed, the parking brake rear cable and conduit tension should be checked.

The ledges of the brake shoes that are positioned against the backing plate support pads should be lubricated with the high-temperature grease recommended by the manufacturer.

7. Inspect wheel cylinder(s) for leakage, operation, and mounting; remove and replace wheel cylinder(s).

During a wheel cylinder inspection check for any leaks that are evident. A slight amount of moisture is acceptable as a small amount of fluid provides for lubrication of the wheel cylinder. Any drops of brake fluid that are evident will indicate a need for replacement. It is not advisable to remove the wheel cylinder dust boots during the inspection processes as dirt and debris could get in the wheel cylinder bore and cause premature wear of the wheel cylinder bore and seals.

Inspect the friction surface of the drum. If the surface does not show signs of wear (the surface is not shiny and worn) this could be evidence that the wheel cylinder is not operating. A pressure test may be made to determine the integrity of the hydraulic system to the wheel cylinder. If the hydraulic system is diagnosed as good, be sure to check the mounting of the wheel cylinder. A bent backing plate/wheel cylinder mount could cause reduced braking. A loosely mounted wheel cylinder could also cause the same condition. If all of the above checks turn out okay, suspect a faulty wheel cylinder.

It is generally suggested that wheel cylinders be replaced in pairs. Dirt and corrosion often form in the bore of the wheel cylinder and lead to wheel cylinder failure. A leaking wheel cylinder will commonly contaminate the shoes and cause the need for shoe replacement as well. When a wheel cylinder is replaced, the hydraulic system will need to be bled. Follow the manufacturer's suggested bleeding sequence when bleeding the hydraulic system. The typical sequence will start with the wheel furthest from the master cylinder. The procedure may also include pulsing all the ABS valves with a scan tool to help release any trapped air bubbles.

8. Install brake shoes and related hardware.

When installing new brake shoes and hardware, the secondary shoe often has a longer lining than the primary shoe and goes on the side of the backing plate pointing toward the rear of the vehicle. The primary shoe faces towards the front of the vehicle, and the primary and secondary shoe return springs are usually not interchangeable. The self-adjuster cable is usually routed on the secondary shoe.

The adjuster must be installed in the proper direction so the star wheel is accessible through the backing plate opening or drum openings to allow for future removal of the shoes and to engage the self-adjuster lever. Be sure to lubricate the pads on the backing plates and any pivot points with appropriate lubricant before installing the shoes and hardware.

9. Pre-adjust brake shoes and parking brake before installing brake drums or drum/hub assemblies and wheel bearings.

Before the brake drum is installed, position a brake adjusting gauge tool inside the brake drum and set the gauge for inside diameter. Next, position the gauge over the brake shoes and adjust the shoes to the gauge using the adjuster wheel. This procedure will pre-adjust the brake shoes and parking brake. This process is also simple and will save time.

On front drum applications and some larger trucks, it will be necessary to adjust the wheel bearing after mounting the drum in place. Follow manufacturer's specifications for correct wheel bearing torque and procedure.

10. Reinstall wheel, torque lug nuts, and make final checks and adjustments.

One area that is a common source of comebacks is improper wheel installation. Wheel studs are designed to retain some elasticity when the wheel is bolted to the vehicle. If the wheels are over torqued this elasticity along with the integrity of the wheel mounting surface are no longer consistent which can result in brake noise, premature rotor warpage or lug nut and stud damage. Recheck brake fluid level, road test, and verify proper brake function. Apply the parking brake and ensure proper parking brake function.

C. Disc Brake Diagnosis and Repair (10 Questions)

1. Diagnose poor stopping, pulling, dragging, or incorrect pedal travel caused by disc brake hydraulic problems; determine needed repairs.

Disc brake systems utilize a disc rotor, caliper assembly, and disc brake pads. The brake pads are positioned on both sides of the brake rotor and are squeezed against the disc rotor by hydraulic pressure acting on the caliper piston or pistons. It was once common to see vehicles built with disc brakes on the front wheels only, however, many cars and trucks today are equipped with four-wheel disc systems.

A cause for dragging brakes could be faulty flex hoses. A restriction in the hose may allow pressure to apply the caliper properly, but not permit the pressure to release. This will keep the piston partially applied and cause the brakes to drag.

Faulty flex hoses can also swell and bulge during brake application. This would cause the driver to feel a springy or soft pedal. Any swollen hoses should be replaced.

Always inspect the disc rotor for signs of bluing or glazing, which is an indication of overheating. Carefully inspect caliper hardware, pins, and slides. Also, closely examine the wear on the disc brake pads for each caliper. Each pad should wear evenly if pressure is being applied the same to each pad. Uneven pad wear may be an indication of caliper with a hydraulic problem.

2. Diagnose poor stopping, noise, pulling, grabbing, dragging, pedal pulsation, or incorrect pedal travel caused by disc brake mechanical problems; determine needed repairs.

There are two brake caliper designs: fixed and floating. Fixed caliper assemblies are rigidly bolted to the spindle or steering knuckle and do not move when the brakes are applied. Fixed caliper assemblies have pistons on both sides that apply hydraulic pressure against the brake rotor.

Floating (or sliding) calipers usually have only one piston that is actuated by hydraulic pressure. The caliper moves or slides as pressure is applied to the caliper piston. This action causes movement or sliding of the caliper and as a result will cause both pads in the caliper assembly to apply pressure to both sides of the brake rotor. These calipers will either float on pins or slide on a machined surface or support bracket. A common problem with floating calipers is tapered brake pad wear. Seized, sticking caliper pins, worn hardware, or corrosion on the support surface can cause the caliper not to float during braking. If one caliper is sticking or seized, it may result in a pull to one side as the brakes are applied.

If the caliper piston is seized, proper brake action will be affected. On vehicles with front disc brakes, a seized caliper piston on one side can cause a pull if the brake pads are not being actuated properly. If the right front caliper is working properly, but the left front caliper piston is seized in its bore, the vehicle will pull to the right when the brakes are applied. If the caliper piston is seized and does not retract when the brake pedal is released, dragging of the brakes will occur. This may cause a pull to one side without applying the brakes and will cause uneven brake wear. Inspect all caliper hardware, which includes anti-rattle clips, caliper pins, caliper slides, and shims. Worn or missing brake caliper hardware may cause noise, chattering, brake squeal, or brake pad rattling.

A pulsation felt through the brake pedal could be the cause of excessive disc rotor runout or loose wheel bearings. Thickness variation (parallelism) will also cause brake pulsation. Sticking or seized calipers can cause brake drag and may damage the rotors due to overheating. This situation may also overheat the brake pads and cause premature, uneven wear.

3. Retract integral parking brake caliper piston(s) according to manufacturers' recommendations.

Many vehicles with four-wheel disc brakes incorporate the parking brake mechanism within the rear brake caliper assembly. These calipers apply the piston hydraulically for normal braking and will apply the caliper piston mechanically when the parking brake is applied.

The caliper piston will extend outward when the parking brake cable acts on a lever connected to the rear of the brake caliper. As the caliper piston extends outward in its bore, the pads clamp the rotor and prevent the vehicle from rolling.

To retract the caliper piston during rear pad replacement on this design caliper, you will need the aid of a special tool called the rear disc brake piston retraction tool. Remove the caliper from its mounting position and remove the brake pads. Using the proper tool, rotate the caliper piston clockwise until it is seated back in its bore. The piston should rotate with some resistance but should retract under pressure applied from the tool. If the piston is seized, replace the caliper. Check parking brake function after reassembly. Adjust the parking brake cable as needed.

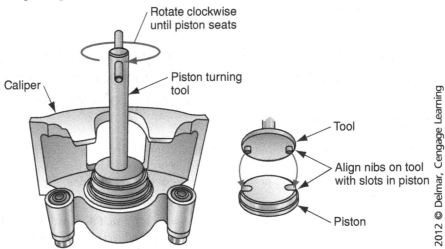

4. Remove caliper assembly from mountings; inspect for leaks and damage to caliper housing.

After removing the caliper, carefully inspect all hardware, mounting contact points, caliper piston boot, and caliper housing. Thoroughly clean the caliper and mounting area. Inspect the dust boot for tears and for leaks from the piston. Also inspect the area where the brake hose connects to the caliper. Check the contact area where the caliper is positioned in the caliper anchor plate. Check for any physical damage to caliper housing or caliper. Any leaks from a caliper are unsafe and the caliper must be replaced.

5. Clean, inspect, and measure caliper mountings and slides/pins for wear and damage.

Inspect the caliper slide pins and sleeve assemblies for corrosion, and inspect all bushings for cuts and nicks. If damage to any of these parts is found, install new parts when the caliper is reinstalled. Do not attempt to polish away corrosion.

The caliper must move freely on the slide or pins with no binding. Inspect the surface area where the caliper is positioned on the anchor plate. All hardware, springs, anti-rattle clips, and pins must be cleaned and inspected. Worn or corroded parts must be replaced.

How the caliper mounts to the anchor or adapter is crucial for proper brake action. A loose caliper fit due to wear may cause brake noise or excessive caliper movement when the brakes are applied. The caliper may rock to one side rather than slide. This action is undesirable and must be corrected. Replacement of the caliper, slides, anchor plate, adapter, or caliper pins may be necessary if excessive wear is determined. All contact surfaces must be lubricated with appropriate brake caliper grease.

6. Remove, clean, and inspect pads and retaining hardware; determine needed repairs, adjustments, and replacements.

A scraping or squeaking noise while braking may be caused by a pad wear sensor contacting the rotor. The scraping of the wear indicator indicates that the brake pads need replacement. If caught early, minimal damage will be done to the rotors. Remember that when resurfacing or replacing the rotors, the axle set must be either replaced together or have the same amount of machining done.

Shop manuals usually specify minimum pad thickness, but the pads can only be measured if the unit is disassembled. Inspect disc brake rotors whenever the pads or calipers are serviced, or when the wheels are rotated or removed for any other work.

After removing the brake pads from the brake caliper, clean and inspect the brake pads, hardware, anti-rattle clips, and springs. Measure disc-pad lining and compare with manufacturer's specification. Check for tapered wear or cracked linings. Weak or broken springs or ant-rattle clips must be replaced. Most caliper slide pins have a coating to protect against corrosion. If the pins are pitted or corroded, do not attempt to clean or polish. The pins must be replaced. Use appropriate brake grease on pins, slides, and brake pad contacting points.

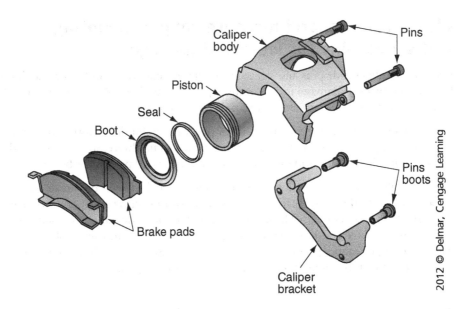

7. Clean caliper assembly; inspect external parts for wear, rust, scoring, and damage; replace any damaged or worn parts; determine the need to repair or replace caliper assembly.

If a brake caliper is to be overhauled, it must be removed from the vehicle. Remove the brake hose and the attached caliper bolt. Carefully remove the caliper from its mounting. Discard the old copper washer used to seal the brake hose. A new sealing washer should always be used during caliper installation. Drain all brake fluid from the caliper and clean the exterior.

Inspect any slide bores for excessive rust or wear that could allow the slide and caliper to move around. Check any mounting surfaces on the caliper that could prevent the caliper from sliding as it was designed. Any defects found should result in caliper replacement.

Before disassembly, make sure the caliper bleeder screw opens. If the bleeder screw is seized and cannot be freed, the caliper will have to be replaced.

The caliper piston can either be removed with air pressure or by using a piston removal tool. If using compressed air, clamp the caliper in a soft-faced vise and place a wooden block between the piston and the caliper body. Apply a small amount of air pressure to the fluid inlet hole to force the piston from the caliper bore. A word of caution: the caliper piston will come out with extreme force, keep hands clear.

After the piston is removed, carefully pry away the dust boot and the caliper seal. Thoroughly clean the caliper bore and inspect for corrosion and pitting. Minor imperfections on the bore surface can be removed by honing. The caliper will need to be replaced if there is severe corrosion or damage. If the caliper bore is honed, check the piston-to-bore clearance. If the clearance is beyond manufacturer's specification, the caliper must be replaced. Inspect the groove where the piston seal is installed in the caliper bore. Carefully inspect groove and bore for pitting and corrosion. Replace the caliper assembly if the seal grove or the caliper bore is damaged.

Carefully examine the caliper piston for damage and wear. On chrome plated pistons, make sure the chrome is not peeling. Also inspect the piston for scoring or pitting. On phenolic pistons, check for cracks, chips, or gouges. Always replace a piston that is damaged. A new dust boot and piston seal will be used for reassembly of the caliper.

Before caliper reassembly, lubricate the caliper seal, piston, and bore with clean brake fluid. Install the seal in the caliper bore first. Then press the piston into the bore until it is fully seated. Using a suitable tool, install the dust boot. A C-clamp may be needed to help fully seat the piston. Extreme care must be used so too much pressure is not applied with the C-clamp.

8. Clean, inspect, and measure rotor with a dial indicator and a micrometer; follow manufacturers' recommendations in determining need to index, machine, or replace the rotor.

The rotor is measured for parallelism or thickness variation; this measurement is usually made at eight locations (every 45 degrees) around the rotor with a micrometer; however, some manufacturers will require measurement in as many as twelve locations. These measurements should be taken near the center of the friction surface. Replace the rotor if the thickness variations exceed the manufacturer's specifications. Two dial indicators set directly across from one another can also be used to determine thickness variation.

Remove the disc brake rotor, clean and inspect for damage. Check for pitting, grooves, heat cracks, or damage on the friction surface of the brake rotor. Blue spots on the rotor face indicate excessive heat. Replace the rotor if there is cracking.

Check all rotors for lateral runout (side-to-side wobble), parallelism (same thickness all the way around), and minimum thickness. Measurements can sometimes be made with the rotor still mounted on the vehicle with the caliper removed.

Before checking lateral runout, if composite rotors are used remove the rotor and clean the back of the rotor and hub surface. This will eliminate any rust or debris that could alter readings. To check runout, mount a dial indicator to a solid surface. Place the indicator near the center of the friction surface and slowly rotate the rotor. See manufacturer's specifications. If runout is excessive, the rotor can sometimes be indexed on the hub to match high spots on the rotor with low spots on the hub. This may allow for runout to be within spec. Some aftermarket companies manufacture shim kits to correct lateral runout. Shims with different thicknesses can be installed between the rotor and hub to correct lateral runout. A micrometer is used to check parallelism. Measure the disc rotor thickness in six or eight spots around the rotor and record the readings. See manufacturer's specifications. Excessive parallelism can cause brake pulsation, brake shudder, and chatter. Also, if the disc rotor thickness is below minimum specification, discard the rotor. Minimum thickness specification is usually stamped into the rotor or can be found in service manuals.

If the disc rotor is not worn beyond specification, truing (or machining) can be performed. Machining the brake rotor involves the use of a brake lathe, which cuts both sides of the rotor in order to produce an even, smooth friction surface. Disc rotors must be machined in pairs to ensure smooth and even braking.

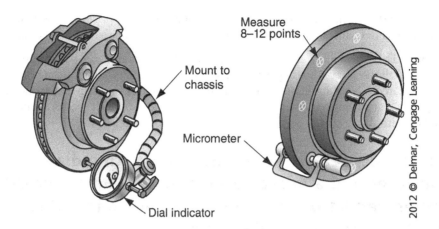

9. Remove and replace rotor.

Disc brake rotors either slide off the axle flange or are mounted in place by the wheel bearing. First, remove the caliper. Do not let the caliper hang by the flex hose; it should be hung up out of the way with a suitable hook or piece of wire. If the rotor is mounted to the axle flange or wheel hub, clean the flange before installing the new rotor. If the rotor is mounted to a spindle and secured with wheel bearings, clean all the grease from the spindle and bearings. Repack the wheel bearings and apply a small amount of wheel bearing grease to the spindle shaft. Following manufacturer's procedure, tighten and torque wheel bearings. Also, always replace the wheel bearing grease seal and use a new cotter pin. Check new rotors for runout, parallelism, and minimum thickness. Clean any oil film from the new rotor prior to installation.

10. Machine rotor, using on-car or off-car method, according to manufacturers' procedures and specifications.

Machining the brake rotor involves the use of a brake lathe, which cuts both sides of the rotor in order to produce an even, smooth friction surface. Remove only enough metal stock to clean up the rotor surface. Never resurface one disc rotor. Disc rotors must be done in pairs to ensure smooth and even braking.

Brake rotors can either be machined on the car or off the car, depending on vehicle design and factory recommendations. An advantage to an on-car brake lathe is that it compensates for any variation in the axle flange, hub, or spindle. When machining rotors off the vehicle, a vibration damper must be positioned around the outside diameter of the rotor or on the rotor faces if the rotor is not vented. A fine grit (180) sanding pad should be used to sand the rotor surface after machining. This helps to knock the tops off of the sharp grooves left by the machining process. Equal amounts of metal must be removed from each side of the rotor.

To determine the approximate amount of metal to be removed, turn on the lathe and bring the cutting bit up against the rotating disc until signs of a slight scratch are visible. Turn off the lathe and reset the depth-of-cut dial indicator to zero. Find the deepest groove on the face of the rotor and move the cutting bit to that point without changing its depth-of-cut position. Now use the depth-of-cut dial to bottom out the tip of the cutter in the deepest groove. The reading on the dial now equals or is slightly less than the amount of material needed to be removed to eliminate all of the grooves in the rotor surface. For the best results with cuts that have a total depth greater than 0.015 inch (0.381 mm), take two or more shallow cuts rather than one very deep cut.

After machining the rotor, measure the rotor for minimum thickness. If the rotor was machined below minimum specification, discard the rotor. Also, give the rotor surface a non-directional finish using the proper lathe attachments. This will ensure proper brake pad to rotor surface contact. A sanding block using 180-grit sandpaper can also be used. Apply the sanding block for thirty seconds on each side of the rotor to achieve a smooth non-directional finish. Rotors must be washed thoroughly after machining. Soapy water works best. Simply submerge the rotor to remove left over debris from machining. Brake clean and rags are not the preferred method because they usually leave behind metal debris that embeds in the new pads and can cause brake squeaks.

11. Install pads, calipers, and related attaching hardware; lubricate components following manufacturers' procedures and specifications; bleed system.

Clean off all caliper supports and sliding surface before installing the caliper assembly and pads. Lubricate sliding surfaces or pins with approved lubricant. It will be necessary to push the caliper piston back into its bore in order to install the brake pads. Open up the caliper bleeder screw first and push the caliper piston back using the appropriate tool. It is not recommended to push the caliper piston back with the bleeder screw closed. The fluid, along with any dirt, can be pushed back through the lines and up to the master cylinder. Install the brake pads using new anti-rattle clips, shims, and other hardware to ensure the pads will sit properly and reduce the chance of squeal and noise. Tighten the caliper guide pins or bolts to the correct toque specification. Check the fluid level in the master cylinder and carefully pump the brake pedal until a firm pedal is felt. Bleed the system following the correct bleeding sequence and road test the vehicle to insure proper brake performance.

12. Adjust calipers with integrated parking brakes according to manufacturers' recommendations.

Some parking brakes on rear disc brake systems utilize a set of brake shoes located inside the hub of the brake rotor. Adjusting the internal parking shoes is performed by rotating the star wheel adjuster with a tool through a slotted opening found in the face of the rotor or an opening found on the splash shield. Rotate the star wheel until correct shoe-to-drum clearance is obtained. Next, apply the parking brake pedal or lever, and remove the slack from the cables, if necessary. There should be no drag at the rear wheels when the parking pedal or lever is released. Apply and release the parking mechanism several times to ensure proper parking brake operation.

Rear disc brake systems that use the caliper piston to apply the parking function only require a cable adjustment.

13. Fill master cylinder to proper level with recommended fluid; inspect caliper for leaks.

Upon completing any brake service, refill the master cylinder with the correct, clean fluid. The brake fluid level on most master cylinders should be .25 inch (6.35 mm) below the top. Many reservoirs are marked indicating full and refill levels. Check with manufacturer for correct fluid type. Actuate the brake pedal to apply hydraulic pressure and inspect the calipers, lines, and hoses for leaks.

14. Reinstall wheel, torque lug nuts, and make final checks and adjustments.

An impact wrench should not be used to tighten wheel lug nuts since the excessive torque from this procedure may distort drums and cause excessive rotor runout.

All lug nuts must be tightened to correct torque specification. Over tightening the lug nuts may damage the brake-rotor hub and cause excessive rotor runout. Before road testing, apply the brake and parking brake to check for correct brake function. Road test the vehicle and ensure proper brake performance.

15. Road test vehicle and burnish/break-in pads according to manufacturers' recommendations.

Road test the vehicle while checking for things like incorrect pedal height, pulling while braking, pulling after breaking, and any abnormal noises. If any of these are present, further diagnosis and repair will be necessary.

Friction material needs to be seated to the matching rotor after pad replacement or machining the rotors. This break-in process is known as burnishing. Burnishing conforms the friction material to the shape of the rotor, which allows for better stopping. The heat generated by this process will also cure the resins and other materials used to make up new pad lining. Failure to burnish newly replaced pads could result in a customer complaint of a hard brake pedal and/or reduced vehicle braking. To seat disc brake pads to the rotor, make approximately 20 complete stops from 30 MPH, or slow down from 50 MPH to 20 MPH the same number of times with light to medium pedal pressure. Allow at least a 30 second cool down period between brake applications to prevent overheating. While performing this procedure, panic/hard stops, which would result in glazed pads, should be avoided. After a thorough burnishing procedure, it is normal to see smoke coming from new brake pads.

D. Power Assist Units Diagnosis and Repair (4 Questions)

1. Test pedal free travel with and without engine running to check power booster operation.

Power assist brakes contain a power booster connected to the master cylinder to reduce the required pedal travel and effort. With a relatively small amount of foot pressure, a great amount of hydraulic pressure can be produced. Most power brake boosters operate by engine vacuum and atmospheric pressure acting on a vacuum diaphragm. With the engine running, vacuum exists on both sides of the booster diaphragm. When the brake pedal is applied, a vacuum port is closed which allows atmospheric pressure to enter one side of diaphragm chamber. This action moves the diaphragm assembly and applies the master cylinder.

With the engine stopped, pump the brake pedal several times, and hold the pedal in the applied position. When the engine is started, the pedal should move slightly downward if the vacuum supply to the brake booster is normal. If the pedal does not move slightly downward, check the vacuum hose and the one-way check valve to the brake booster.

To check the vacuum brake booster for air tightness, operate the engine for two minutes and then shut off the engine. Pump the brake pedal several times with normal braking pressure. If the brake booster is operating normally, the pedal should go down normally on the first brake application. The pedal should gradually become higher and harder to depress with each pedal application.

2. Check vacuum supply (manifold or auxiliary pump) to vacuum-type power booster.

Vacuum assisted power brakes rely on a steady, adequate supply of manifold vacuum. Internal engine problems can affect power assist if manifold vacuum is lower than normal. Any vacuum leaks will also affect proper booster operation. A vacuum leak in the booster assembly can also occur. Low vacuum or a leaking booster will cause a hard brake pedal with little or no power assist. Engine vacuum should be 17–20 in/Hg @ sea level at idle.

To check for proper vacuum, disconnect the booster hose from the manifold and connect a vacuum gauge. On vehicles with auxiliary pumps, test for vacuum at the pump with the engine running. Vacuum should be 15–22 in/Hg @ sea level. Compare that reading right at the booster to make sure the booster is being supplied with the same vacuum.

3. Inspect the vacuum-type power booster unit for vacuum leaks and proper operation; inspect the check valve for proper operation; repair, adjust, or replace parts as necessary.

To check a one-way check valve in the brake booster vacuum hose, operate the engine at 2,000 rpm and then allow the engine to idle. Shut the engine off and wait 90 seconds. Pump the brake pedal five or six times. The first two pedal applications should be power assisted. If the first two brake applications are not power assisted, the one-way check valve is defective in the brake booster vacuum hose.

Insufficient manifold vacuum, leaking or collapsed vacuum lines, punctured diaphragms, or leaky piston seals can all result in weak power-unit operation. A steady hiss when the brake pedal is held down indicates a leak that can cause poor power-unit operation. Hard brake pedal is usually the first signal that the unit is on the way to complete failure.

Connect a vacuum gauge with a T-connection in the hose between the one-way check valve and the brake booster. With the engine idling, the vacuum should be 17 to 20 in/Hg @ sea level (44 to 30.4 kPa absolute). If the vacuum is low, connect the vacuum gauge directly to the intake manifold and check this reading against the manufacturer's specifications.

4. Inspect and test hydro-boost system and accumulator for leaks and proper operation; repair or replace parts as necessary; refill system.

The hydraulic power brake booster system (hydro-boost) is used on some cars and trucks. Hydro-boost systems use hydraulic pressure developed from the power steering pump to actuate the master cylinder. Some vehicles use belt driven pumps, others use electric driven pumps. A hydro-boost system consists of a pump, booster assembly, master cylinder, and an accumulator.

As the brake pedal is depressed, the booster pushrod and piston are moved forward. This action causes the spool valve to move and allow fluid flow behind the power piston. As hydraulic pressure builds, it actuates the master cylinder.

The accumulator is used as a back-up should the hydraulic power source fail. Accumulators are either spring-loaded or contain gas under pressure. The accumulator is filled with hydraulic fluid and is pressurized when the brakes are applied. If the engine stalls or a failure occurs in the power steering pump, the accumulator will have an adequate amount of pressurized fluid to provide one to three power assisted brake applications. To test the accumulator, run engine and then shut off. Wait five to ten minutes and test brake assist. If the accumulator is working correctly, there will be two to three assisted applications.

The inspection and testing of a hydro-boost system must include a complete inspection of the power steering pump, belt, lines, hose connections, and the hydro-boost unit. Any leaks in the power steering system will affect hydro-boost operation. Aerated power steering fluid will also cause the hydro-boost system not to function correctly. Low fluid level can cause a moan with a vibration in the pedal and steering column, usually experienced during parking or low speed maneuvers. It's important to remember that the pressure needed to actuate the hydro-boost unit and master cylinder originates from the power steering pump assembly, which is usually operated by a belt, driven by the engine. A slipping belt or lower pump pressure will affect hydro-boost operation.

Bleeding the power steering system will be necessary after repairs are made to the power steering system or to the hydro-boost assembly. To bleed the system, perform the following procedure:

- Start the engine and apply the brake pedal several times while turning the steering wheel from stop-to-stop. Turn the engine off and apply the brake pedal several more times to deplete accumulator pressure. Check fluid level and add if needed. If the fluid is foamy, let the vehicle stand for a few minutes and recheck fluid level. It may be necessary to repeat the process until all air is purged from the system.
- With the wheels off the ground and the engine stopped, turn the steering wheel from stop-to-stop. Check fluid level and add if needed. Lower the vehicle.
- Repeat the first step: start the engine and apply the brake pedal several times while turning the steering wheel from stop-to-stop. Turn the engine off and apply the brake pedal several more times to deplete accumulator pressure. Check fluid level and add if needed. If the fluid is foamy, let the vehicle stand for a few minutes and recheck fluid level. It may be necessary to repeat the process until all air is purged from the system.

Vehicles with hydraulically assisted power brakes will occasionally develop a problem where power steering fluid gets forced into the brake fluid. If this happens the entire hydraulic brake must be repaired. All rubber components which have come into contact with the oil/brake fluid mixture will swell and must be replaced. All metal components must be thoroughly flushed to remove the contamination.

E. Miscellaneous Systems (Pedal Linkage, Wheel Bearings, Parking Brakes, Electrical, etc.) Diagnosis and Repair (7 Questions)

1. Diagnose wheel bearing noises, wheel shimmy, and vibration problems; determine needed repairs.

The wheel bearings perform a major role in effective brake function and steering performance. There are two different types of wheel bearing designs used on modern automobiles: the adjustable tapered roller bearing and the non-adjustable sealed roller or ball bearing.

Worn wheel bearings can cause a growling noise and vibration when the vehicle is driven. Checking for loose or worn wheel bearings should always be part of a complete brake inspection. Wheel bearings that are worn or improperly adjusted can cause poor brake performance, poor steering, and rapid wheel bearing wear.

Tapered wheel bearings can be disassembled, cleaned, re-greased, and adjusted. Sealed roller or ball bearing wheel bearings cannot be serviced and are replaced as a unit.

2. Remove, clean, inspect, repack wheel bearings, or replace wheel bearings and races; replace seals; replace hub and bearing assemblies; adjust wheel/hub bearings according to manufacturers' specifications.

Tapered roller bearings are generally used on non-drive axles. The wheel bearings are mounted between a hub and a fixed spindle. To gain access to tapered wheel bearings, remove the wheel, brake rotor or drum, dust cap, cotter pin, and spindle nut, and remove the outer wheel bearing. Remove the hub/rotor or hub/drum assembly. Pry the wheel bearing seal out and remove the inner wheel bearing. Thoroughly clean the wheel bearing, hub assembly, spindle shaft, and races. Carefully inspect the wheel bearings, spindle, and races for signs of wear. Discard bearings showing any signs of wear, chipping, bluing, or galling. Repack the wheel bearings with high temperature grease. Never repack a wheel bearing without first removing all the old grease. Insert the inner bearing into the hub and lightly lubricate the new wheel seal with grease. Tap the seal into place using a seal driver. Carefully install the hub/rotor or hub/drum assembly onto the spindle. Install the outer wheel bearing, washer, and spindle nut. Always reinstall a bearing in the same race.

If the wheel bearings need replacing it will be necessary to replace the bearing races also. Remove the bearing race from the hub using a brass drift or appropriate puller. Drive the new races into the hub until fully seated using a bearing race installer. If using a drift punch, tap the races a little at a time, moving the punch around the race to avoid cocking. Use a soft steel drift, never a hardened punch.

It is very important to properly adjust tapered wheel bearings. Always check with manufacturer's recommended procedure for the specific vehicle being serviced. There are two widely used methods for adjusting wheel bearings: the torque wrench method and the dial indicator method.

In the torque wrench method, rotate the wheel in the direction of tightening while the spindled nut is tightened to the specified torque. This initial torque setting seats the bearings in the bearing race. The nut is then loosened until it can be rotated by hand. The nut is then torqued, this time to lower specified value. Back the nut, if necessary, to install the cotter pin and lightly tap on the dust cap.

To adjust the wheel bearings using the dial indicator method, start by tightening the spindle nut while spinning the wheel, to fully seat the bearings. Loosen the spindle nut until it can be rotated by hand. Mount a dial indicator so the indicator point makes contact with the machined outside face of the hub. Firmly grasp the sides of the rotor or tire, and pull in and out. Adjust the spindle nut until the end play is within manufacturer's specification. Typical end play ranges from .001 inch to .005 inch. Install the cotter pin and dust cap.

Sealed ball and roller bearings are not serviceable and must be replaced when they are defective or have damaged grease seals. Removing this type of bearing involves pressing the bearing from the bearing hub out of the spindle or knuckle assembly. Carefully inspect the bearing, hub, and spindle assembly for wear. Press the new bearing into the spindle/knuckle assembly and torque the axle nut following the manufacturer's procedure.

3. Check parking brake system; inspect cables and parts for wear, rust, and corrosion; clean or replace parts as necessary; lubricate assembly.

The parking brake is a mechanical system integrated with the hydraulic brake system. It is intended to prevent the vehicle from rolling while it is parked. The rear wheels are held in place by a series of cables and linkages when either a parking brake pedal or hand lever is applied.

The parking brake cables must move freely when applied. Inspect the cables and all parking brake components for wear, binding, rust, and corrosion build up. Check for cable wear at points of contact. A frayed cable must be replaced. Inspect the parking brake or lever assembly for binding and malfunction. Make sure the parking brake ratchets into place and locks into position. Check all the attaching cable hardware inside the brake drum. For disc brakes, inspect the hardware and lever at the rear of the caliper for wear and malfunction. Lubricate the cables and pivot areas as needed. Some cables are coated and cannot be lubricated.

4. Adjust parking brake assembly; check operation.

To check operation of the parking brake, raise the vehicle off the ground and apply the parking brake. The rear wheels should be locked in position. If the wheels can be rotated in the forward direction by hand with the parking brake applied, the parking brake may require adjustment. Before any adjustment is made, always check the condition of the brakes and drum brake adjustment. Worn rear brake shoes or brakes out of adjustment will affect the operation of the parking brake. If the parking brake is incorporated with the rear disc caliper, check the calipers for proper operation.

To adjust the parking brake assembly, raise the vehicle off the ground and apply the parking brake two or three clicks. There should be no slack in the cables and a slight drag felt at the rear wheels. If necessary, adjust the equalizer nut until all the slack is removed from the cables and a slight drag felt at the rear wheel. Release the parking brake.
The parking brake should release and the wheels should turn freely. Next apply the parking brake about six clicks to see if the rear brakes lock. Make a final adjustment if needed. Check the service manual for specific procedure and number of required clicks to lock the rear wheel.

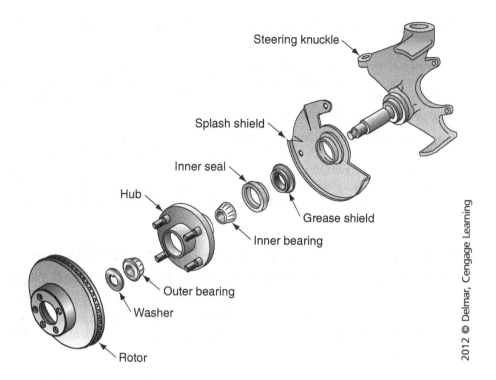

5. Test the parking brake indicator light, switch, and wiring.

Battery voltage is applied to the brake warning indicator light when the ignition is in the run mode, bulb test mode, or start mode. The parking brake switch supplies the ground for the warning light. The ignition switch supplies an additional ground for the brake warning light in the start or bulb test mode to provide a means to check proper bulb function.

A continually illuminated brake warning light may be caused by a grounded wire to the parking brake switch, a continually closed parking brake switch, or low fluid level in one section of the master cylinder reservoir. If an open circuit occurs in the wire to the parking brake switch, the brake warning light is not illuminated when the parking brake is applied.

When the parking brake pedal is applied, it closes a switch which completes an electric circuit to the brake indicator light in the instrument panel. The parking brake applied indicator light will then light when the ignition is turned on. The light goes out when the parking brake is released or the ignition is turned off. In some vehicles, this same indicator light may be used to alert the driver to problems in the antilock brake system.

6. Test, adjust, repair, or replace brake stop light switch, lamps, and related circuits.

The brake light switch can be adjusted on some vehicles. The brake lights should operate when the brake pedal is depressed 0.25 inch (6.35 mm). Battery power is supplied to one side of the switch, which is normally open. When the brake pedal is depressed, the switch closes, supplying battery power to the brake light bulbs.

If the brake lights do not operate, check for battery power to the brake light switch. If there is no battery power, check the fuse or check for an open wire to the brake light switch. If battery power is present at the switch, depress the brake pedal and check for battery power on the other side of the switch, which feeds battery power to the brake light bulbs. No power on the other side of the switch with the brake pedal depressed indicates a faulty switch.

If the switch is operational but the brake lights do not operate, check for an open wire either on the battery feed side from the switch or ground side of the wiring harness at the brake light bulbs. Always consult the service manual for the specific wiring diagram when diagnosing the brake light circuit.

Some brake light switches contain multiple contacts to control cruise control function and torque converter clutch operation. Also, the brake light circuit on many vehicles is integrated with the turn signal switch or combination switch in the steering column. A defective turn signal switch or combination switch may affect the brake light circuit. On vehicles equipped with antilock brakes, the brake switch is also an input to the brake control module.

7. Inspect and test brake pedal linkage for binding, looseness, and adjustment; determine needed repairs.

Part of brake inspection includes checking the brake pedal and linkage for wear and proper function. Check for binding or worn bushings. Worn bushings or loose linkage may affect brake performance. If the vehicle was involved in an accident, check for a bent brake pedal.

F. Electronic Brake Control (EBC) Systems: Antilock Brake System (ABS), Traction Control Systems (TCS), and Electronic Stability Control System (ESC) Diagnosis and Repair (7 Questions)

1. Follow manufacturers' service and safety precautions when inspecting, testing, and servicing electronic brake control system, hydraulic, electrical, and mechanical components.

Many vehicles today are equipped with antilock brakes (ABS) systems, traction control systems (TCS), and electronic stability control (ESC) systems. Although vehicle designs may differ, all antilock systems essentially function the same way. The ABS computer or control module monitors wheel speed, vehicle speed, and other vehicle functions. The wheel speed sensors send vital information to the ABS control module. When the control module senses that a wheel is about to lockup, it will regulate hydraulic pressure to the wheels in order to prevent lockup. During a panic stop situation, the antilock brake system allows the driver to maintain directional control while providing maximum braking performance. Traction control systems use the same sensors as ABS to help reduce wheel spin. The traction control system can command the engine to reduce torque as well as apply a brake to a spinning wheel in order to help the vehicle regain traction. Electronic stability control systems incorporate a yaw rate sensor which can detect slip angle. When the ESC system detects lateral acceleration outside of program limits the brakes can be selectively applied at each wheel to help bring the vehicle back under control.

Most of the services done to the brakes of a vehicle equipped with these systems are identical to those in a conventional brake system. There are, however, some important differences. One of these is the bleeding of the brake system. Always refer to the appropriate procedures in the service manual before attempting to service the brakes on an electronic brake control equipped vehicle. Before servicing an electronic brake control system, it is important that you understand the basics of electrical and electronic troubleshooting. Without this basic understanding, it will be difficult to follow the diagnostic procedures given in most service manuals.

Servicing electronically controlled brakes require following different safety precautions. Always refer to the service manual and review safety procedures for the specific vehicle being serviced.

These common safety guidelines should be observed:

1. Always wear safety glasses.
2. Never open a bleeder valve or loosen a brake line on an integral ABS system when the accumulator is pressurized.
3. Remove the pressure from the accumulator by pumping the brake pedal 40 times with the ignition key in the off position.
4. Never connect or disconnect any electrical connector or component with the key in the on position. This may damage the ABS controller.
5. Never strike or tap on wheel speed sensor components. Striking these components may de-magnetize the wheel speed reluctor and will affect the signal strength of the sensor.
6. Never mismatch tire sizes.
7. DOT 5 fluid should never be used in ABS systems.

2. Diagnose poor stopping, wheel lockup, false activation, pedal feel and travel, pedal pulsation, and noise concerns associated with the electronic brake control system; determine needed repairs.

The clicking noise during initial driving is a result of the ABS computer self-test mode in which the computer momentarily energizes the solenoid(s) in the ABS system. On many systems, pedal pulsations are normal during ABS function; however, pedal pulsation during a normal stop when the ABS function is not operating may be caused by out-of-round drums or rotors with excessive runout.

Vehicles equipped with antilock brakes will have different braking characteristics during extreme braking than vehicles without ABS. During normal braking conditions, the antilock portion of the brake system does not function. However, wheel speed sensors continuously send information to the ABS control module. When wheel lockup begins to take place, the control module will modulate hydraulic pressure through a series of solenoids to prevent wheel lockup and maintain steering ability. Correct tire size plays a critical role in antilock operation. Differently sized tires can affect ABS operation and may cause wheel lockup. During the ABS function process, the driver can misinterpret normal ABS operation as a brake problem. A pulsation felt through the brake pedal during ABS operation is normal. Also, a whining or ratcheting noise can be heard as the solenoids are modulating hydraulic pressure to the wheels. Some systems use an antilock pump assembly, which will initiate from time to time to maintain pressure in the accumulator. The driver may hear an audible whine from the pump. If the pump runs continuously or is activated for longer periods of time, this may indicate a faulty accumulator.

If no pulsation is felt under hard braking at speeds above 10 mph, suspect an antilock malfunction. Below 10 mph, the antilock system is generally disabled.

False activation of the electronic brake control system can sometimes be attributed to a sensor to toner wheel gap that is too large. Some sensors can be adjusted while others may have other corrective actions. Damage to the sensor or other suspension components can lead to excessive gap. Sensors that bolt on to a sealed hub assembly can have rust build-up between the sensor and mounting surface. Cleaning the mounting surface should restore proper sensor to toner wheel clearance.

3. Observe electronic brake control system indicator light(s) at startup and during road test; determine if further diagnosis is needed.

All electronic brake control systems have some sort of self-test. This test is activated each time the ignition switch is turned on. You should begin all diagnostics with this simple test. To perform a typical EBC self-check sequence, place the ignition switch in the START position while observing both the red system light and the electronic brake control indicator lights. Both lights should turn on. Start the vehicle. The red brake system light should turn off. With the ignition switch in the RUN position, the electronic brake control module performs a preliminary self-check on the EBC electrical system. During the self-check the lights typically remain on (three to six seconds) and then should turn off. If a malfunction is detected during the test, the lights will remain on and the system will shut down.

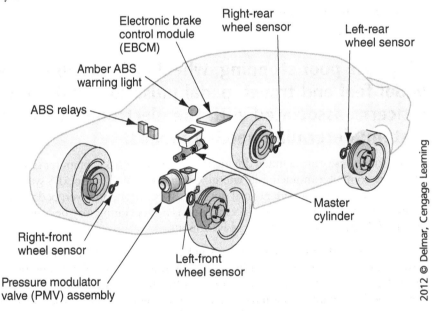

4. Diagnose electronic brake control system, electronic control(s), components, and circuits using on-board diagnosis and/or recommended test equipment; determine needed repairs.

Most modern electronic brake control systems have the ability to store an error message, called a fault code, should a malfunction occur. By using a scan tool, the fault code can be retrieved from the computer memory. The fault code will direct you to troubleshooting procedures and aid in diagnosing the malfunction. On some systems, identifying the fault code is accomplished by putting the system in diagnostic mode and counting the number of times the indicator lamp flashes. Always consult the appropriate service manual for specific diagnostic procedures.

Always begin your diagnosis with a complete visual inspection of the entire brake system and electronic brake control system. Check fluid level, harness connectors, and wiring to components. If a fault code is stored in the EBC controller memory, follow the appropriate diagnostic procedure to diagnose the problem. Many systems allow the technician to view sensor and component data through a scan tool. Vital information, such as wheel speed sensor operation, can be viewed, compared, and analyzed. The use of a scan tool can greatly reduce the diagnostic time needed to solve tough ABS problems.

5. Bleed and/or flush the electronic brake control hydraulic system following manufacturers' procedures.

Bleeding procedures for electronic brake control systems will vary depending on the vehicle. Different EBC systems will require different bleeding methods. Some vehicles will require the use of a bi-directional scan tool to move solenoids or other components in a particular position. Many non-integral EBC systems can be bled using traditional methods. On most integral systems, a fully charged accumulator and special bleeding equipment is needed. Always refer to the service manual for specific recommended bleeding procedures.

6. Remove and install electronic brake control system components following manufacturers' procedures and specifications; observe proper placement of components and routing of wiring harness.

Some electronic brake control systems utilize a pump, motor, hydraulic actuator, and accumulator with the master cylinder in one single unit (called integral EBC). Other systems connect the EBC hydraulics to the master cylinder and other components by means of high-pressure hoses. Most EBC components, such as wheel speed sensors, control modules, and hydraulic actuators are not serviceable and should be replaced if found defective.

When connecting and disconnecting any EBC components, make sure that the ignition key is in the off position. Some manufacturers advise removing the battery cables. Always make sure the ignition key is off and brake system depressurized before starting any repairs. Pumping the brake pedal at least 25 to 40 times will depressurize most antilock brake systems.

All wiring to any EBC component must be routed the exact same way it was removed. Do not alter the position, routing, or mounting of the harness.

7. Test, diagnose, and service electronic brake control system sensors (such as speed, yaw, steering angle, brake pedal position, etc.) and circuits following manufacturers' recommended procedures (includes output signal, resistance, shorts to voltage/ground, and frequency data).

The wheel speed sensors continuously send signals to the EBC controller. The typical wheel speed sensor consists of a toothed ring (or tone wheel), made from ferrous metal, and a permanent magnet sensor. As the wheel turns, the wheel sensor produces a varying ac voltage based on the changing magnetic field. The strength of the signal is dependent upon the speed of wheel rotation. The ac voltage will alternate from a maximum positive voltage to a maximum minimum voltage. This cycle is repeated many times and is known as a sine wave. By counting the number of cycles occurring each second the ABS controller can determine the frequency of the wheel speed sensor. The controller can use this information and compare the frequency of one wheel speed sensor against another. One wheel speed sensor decelerating faster than another indicates an impending wheel lockup.

When testing a wheel speed sensor, start with a visual inspection of the sensor and wiring. Debris trapped around the sensor or ring can affect signal quality and strength. Also, if possible, check the air gap between the sensor and the toothed ring. A gap too wide may cause a weak signal or erratic readings. Inspect the toothed ring for debris, chips or cracks. Some sensors are mounted in the differential with the tone ring located on the ring gear. Metal particles can accumulate around the sensor from bearing and gear wear.

An ohmmeter can be used to check the resistance of the sensor. Compare the readings with the service manual. Probably the most accurate way to test a wheel speed sensor is with a digital storage scope. A scope can test the quality of the signal and spot any intermittent glitches in the sensor signal.

The wheel speed sensor harness can also be a source of problems. The harness must be able to cleanly send information to the EBC controller. A misrouted sensor harness too close to a heat source or magnetic field can result in false information being sent to the controller. Corrosion in the harness or in a harness connecter can cause unwanted resistance in the wiring and a diminished wheel speed sensor signal. Always check the harness routing and check the continuity and quality of the harness wiring.

Recently some manufacturers have started using Hall-effect wheel speed sensors. These sensors have the ability to detect the direction the wheel is turning and are useful when a vehicle is equipped with a hill holding option. These sensors can be tested with a scan tool.

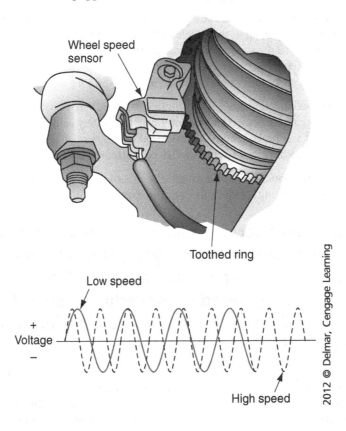

8. Diagnose electronic brake control system braking concerns caused by vehicle modifications (wheel/tire size, curb height, final drive ratio, etc.) and other vehicle mechanical and electrical/electronic modifications (communication, security, radio, etc.).

All EBC computers are specifically programmed for the particular vehicle they are controlling. All aspects of the vehicle are taken into account: vehicle weight, transmission type, tire size, differential, suspension characteristics, electronic devices, and more.

Changing the tires and wheels to a size not recommended by the manufacturer can greatly affect EBC system operation. The EBC controller is programmed to look for and expects to see specific information being sent by the wheel speed sensors. By changing the tire and wheel size, the signal will be altered and may cause a malfunction. It should be noted that many late model trucks have the ability to change their programming for optional tire sizes. As with all things relating to vehicles every rule has an exception.

Any modification to the vehicle may affect EBC operation. If an electronic device, such as radio or an alarm, is installed, make sure the wiring harness does not interfere with EBC harness. When replacing differentials and transmissions always confirm the correct final drive ratio.

9. Repair wiring harness and connectors following manufacturers' procedures.

Always handle the EBC harness with care when testing, repairing, or servicing components. Never probe through the harness insulation and be very careful when back probing a connector. Follow all guidelines prescribed in the service manual for testing components, connectors, and wiring. If a harness connector is faulty, do not attempt to repair. If a replacement connector is not available, replace the harness. If a section of harness wiring is damaged, it is recommended not to repair the wire, but to replace the damaged harness.

SECTION 5

Sample Preparation Exams

INTRODUCTION

Included in this section is a series of six individual preparation exams that you can use to help determine your overall readiness to successfully pass the Brakes (A5) ASE certification exam. Located in Section 7 of this book you will find blank answer sheet forms you can use to designate your answers to each of the preparation exams. Using these blank forms will allow you to attempt each of the six individual exams multiple times without risk of viewing your prior responses.

Upon completion of each preparation exam, you can determine your exam score using the answer keys and explanations located in Section 6 of this book. Included in the explanation for each question is the specific task area being assessed by that individual question. This additional reference information may prove useful if you need to refer back to the task list located in Section 4 for additional support.

PREPARATION EXAM 1

1. A drum is chattering on the brake lathe. This could be the result of:
   A. Failure to install the dampening belt.
   B. Cutting too slowly.
   C. Dull cutting blades.
   D. Cutting speed too fast.

2. A vehicle equipped with a hydro-boost brake system has a hard pedal with little braking action. Which of the following could be the cause?
   A. Low brake fluid
   B. Air in the brake fluid
   C. Low power steering fluid
   D. An overtightened power steering belt

3. Which of the following could cause excessive noise while braking?
   A. Worn wheel cylinders
   B. Worn caliper pistons
   C. Seized caliper slides
   D. Missing clips

4. A vehicle equipped with a "hill holding" option has an ABS code. Which tool would most likely be used to check the wheel speed sensor?

   A. AC volt meter
   B. Ohm meter
   C. Scan tool
   D. Ammeter

5. The brake system has a soft, sinking pedal with the engine running. The brake fluid level is at the correct height. Which of the following could be the cause?

   A. A restricted vacuum hose
   B. A leaking master cylinder
   C. A bypassing master cylinder
   D. A stuck open vacuum check valve

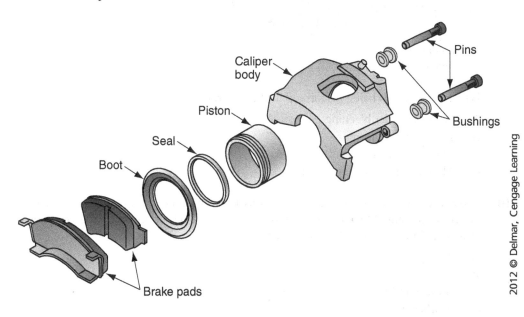

6. Refer to the figure above. Technician A says the pins should be driven in with a brass hammer. Technician B says the pins should be lubricated with brake fluid. Who is correct?

   A. A
   B. B
   C. Both A and B
   D. Neither A nor B

7. Technician A says a hydro-boost system can use the power steering pump to operate the booster. Technician B says a hydro-boost system can use engine vacuum to operate the booster. Who is correct?

   A. A
   B. B
   C. Both A and B
   D. Neither A nor B

8. A disc brake caliper is dragging. All of the following could be the cause EXCEPT:
   A. Binding pins.
   B. Sticking piston.
   C. Binding caliper.
   D. Worn brake pads.

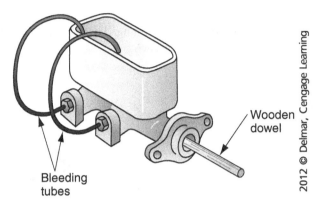

9. Which operation is being performed in the illustration above?
   A. Wheel cylinder bleeding
   B. Brake caliper bleeding
   C. Master cylinder bleeding
   D. Master cylinder pressure test

10. The rear wheels of a front wheel drive car lockup during heavy braking. Technician A says this could be caused by a faulty proportioning valve. Technician B says this could be caused by a faulty brake booster. Who is correct?
    A. A only
    B. B only
    C. Both A and B
    D. Neither A nor B

11. There is a loud growling noise only when the vehicle is steered to the right while driving. Which of the following could be the cause?
    A. Worn left front brakes
    B. Worn wheel bearing
    C. Worn right front brakes
    D. Sticking brake caliper

12. A spongy pedal can be the result of:
    A. Leaking wheel cylinders.
    B. Broken return springs.
    C. Contaminated linings.
    D. Seized wheel cylinder pistons.

13. A vehicle has a wheel speed sensor diagnostic trouble code. Which of the following is the LEAST LIKELY cause?

    A. Faulty wheel speed sensor
    B. Cracked wheel speed sensor reluctor ring
    C. A tire with the incorrect height
    D. A tire with the incorrect width

14. A vehicle has a squeak only when applying the brakes. Which of the following could be the cause?

    A. Aerated brake fluid
    B. Water contaminated brake fluid
    C. Worn brake pads
    D. Worn wheel bearings

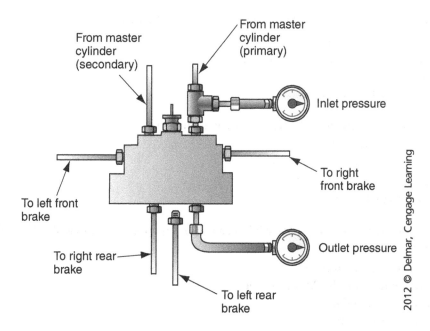

15. During a moderate brake pedal application, with the engine running, which of the following would be considered a normal pressure reading on the inlet pressure gauge shown in the figure above?

    A. 100 psi
    B. 900 psi
    C. 3000 psi
    D. 5000 psi

16. Which of the following would be the most normal end-play specification for a set of tapered roller front wheel bearings?

    A. 0.0001"–0.0005" (0.00254 mm–0.0127 mm)
    B. 0.0010"–0.0050" (0.0254 mm–0.127 mm)
    C. 0.0100"–0.0500" (0.254 mm–1.27 mm)
    D. 0.1000"–0.5000" (2.54 mm–12.7 mm)

17. Brake chatter can be caused by:
    A. Seized star-wheel adjuster.
    B. Seized wheel cylinder.
    C. Contaminated linings.
    D. Switched primary and secondary shoes.

18. Technician A says an analog wheel speed sensor can detect which direction the wheel is turning. Technician B says a digital wheel speed sensor can detect which direction the wheel is turning. Who is correct?
    A. A
    B. B
    C. Both A and B
    D. Neither A nor B

19. All of the following are the result of brake pad burnishing EXCEPT:
    A. Forms the new pads to the rotor.
    B. Cures the materials in the friction material.
    C. Provides for better braking.
    D. Cures the metal in the rotor.

20. When bench bleeding a master cylinder which statement is true?
    A. New brake fluid should be used.
    B. Brake cleaning fluid should be used.
    C. Mineral spirits should be used.
    D. Mineral oil should be used.

21. The technician is preparing to do a brake job on a vehicle equipped with integral ABS. Which of the following should be done?
    A. Pump the pedal 25–40 times
    B. Remove the fuses for the ABS controller
    C. Apply the parking brake
    D. Remove the fuse for the ignition system

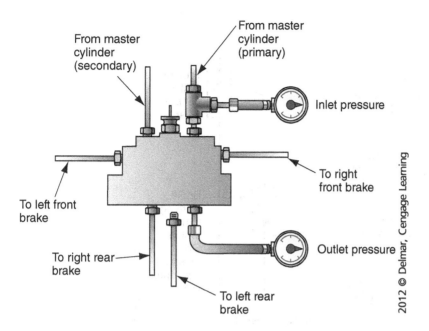

22. Refer to the illustration above. Technician A says pressure is being tested at the antilock braking system (ABS) modulator valve. Technician B says pressure is being tested at the combination valve. Who is correct?

    A. A only
    B. B only
    C. Both A and B
    D. Neither A nor B

23. A customer has a noise concern coming from the rear of the vehicle. All of the following could be the cause EXCEPT:

    A. Worn rear wheel bearings.
    B. Rear tires out of balance.
    C. Dragging brake shoes.
    D. Binding parking brake cable.

24. A vehicle comes back into the shop with a brake pulsation after having the rotors replaced. Which of the following is the most likely cause?

    A. Over torqued lug nuts
    B. Under torqued lug nuts
    C. Sticking slides
    D. Incorrect brake fluid installed

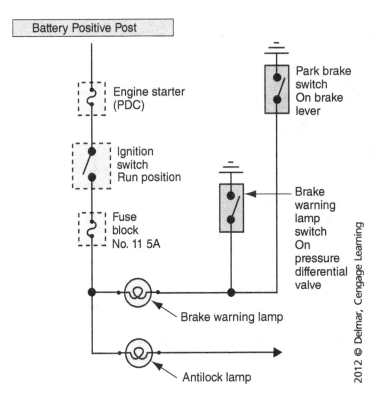

25. The brake warning lamp illustrated in the figure above does not illuminate when the parking brake is applied. It operates normally otherwise. Which of the following could be the cause?

    A. An open ground at the parking brake switch
    B. An open ground at the master cylinder
    C. An open fuse 11
    D. An open engine starter fuse

26. A customer is concerned that on dry pavement, their vehicle will occasionally have a rapid brake pulsation similar to the feel of an ABS stop. Which of the following is the most likely cause?

    A. Faulty wheel speed sensor
    B. Faulty master cylinder
    C. Faulty ABS control module
    D. Faulty brake booster

27. The caliper piston seal is leaking. Which of the following would most likely result?

    A. The ABS light would start to flash.
    B. The traction control light would start to flash.
    C. The brake pedal would become very hard.
    D. The brake pedal would become very soft.

28. A vehicle pulls right while braking. The most likely cause would be:

    A. Seized left caliper slides.
    B. Faulty metering valve.
    C. Seized right rear wheel cylinder.
    D. Faulty master cylinder.

29. The pressure differential valve lights the red brake warning light when:
    A. The parking brake pedal is depressed.
    B. There is an ABS code.
    C. There is a pressure imbalance.
    D. The brake fluid is low.

30. A vehicle equipped with a hydro-boost brake system is being diagnosed for a high, hard pedal. Which of the following tools could be used?
    A. A vacuum gauge
    B. A black light
    C. A power steering pressure tester
    D. A vacuum pump

31. A vehicle has a low, firm brake pedal. Which of the following would most likely be the cause?
    A. Improperly adjusted brake shoes
    B. Leaking wheel cylinders
    C. Excessive rotor runout
    D. Out of round drums

32. The brake fluid reservoir is low on fluid. Which of the following is the most likely cause?
    A. Worn pads
    B. Worn shoes
    C. Bypassing master cylinder
    D. Leaking vacuum check valve

33. A customer says their car makes more noise on asphalt than on concrete road surfaces. Which of the following could be the cause?
    A. Front wheel bearing wear
    B. Rear wheel bearing wear
    C. Aggressive tire tread patterns
    D. Worn brake pads

34. Which of the following would most likely cause a brake fluid leak on a front disc brake caliper assembly?
    A. Piston seal
    B. Piston boot
    C. Caliper slide pins
    D. Caliper slide bushings

35. All of the following could cause a brake pedal pulsation EXCEPT:
    A. Thin brake pad linings.
    B. Excessive lug nut torque.
    C. Excessive rotor thickness variation.
    D. Rust build-up between the hub and rotor.

36. There is a loud growling noise while driving. The noise gets louder when vehicle speed increases. Which of the following could be the cause?
    A. Worn rear brake shoes
    B. Worn wheel bearing
    C. Worn ball joint
    D. Worn front brake pads

37. Which of the following would be the correct way to handle brake fluid?
    A. Because brake fluid does not have an expiration date, purchase in the largest containers possible for economy.
    B. Store in a container vented to the atmosphere.
    C. Store in small, tightly sealed containers.
    D. Store in a temperature controlled environment, preferably refrigerated.

38. Technician A says rust can cause a faulty wheel speed sensor signal. Technician B says a defective brake switch can cause a faulty wheel speed signal. Who is correct?
    A. A
    B. B
    C. Both A and B
    D. Neither A nor B

39. A vehicle equipped with a hydro-boost brake system has a low, soft pedal with little braking action. Which of the following could be the cause?
    A. Weak power steering pump
    B. Air in the brake fluid
    C. Air in the power steering fluid
    D. Low intake manifold vacuum

40. A set of tapered roller bearings are being serviced. Which of the following is true concerning the correct tightening procedure?
    A. Tighten the nut to 75 ft lbs.
    B. Tighten the nut to spec and back off one turn.
    C. The final check should be made with a dial indicator.
    D. The wheels/tires must be on the ground.

41. The brake system has a low, soft pedal with the engine running. The brake fluid is low in the primary reservoir. Which of the following could be the cause?
    A. A restricted vacuum hose
    B. A leaking master cylinder
    C. A bypassing master cylinder
    D. A stuck open vacuum check valve

42. A brake line needs replaced. Technician A says that a new section of line should be installed with compression fittings. Technician B says a new section of line should be installed using an ISO or double flare. Who is correct?
    A. A only
    B. B only
    C. Both A and B
    D. Neither A nor B

43. Which of the following is a typical input to the ABS control module?

    A. Brake light switch
    B. Pressure differential switch
    C. PRNDL switch
    D. Power steering pressure switch

44. Technician A says that brake drag can be caused by worn drums. Technician B says that brake drag can be caused by a seized star-wheel adjuster. Who is correct?

    A. A only
    B. B only
    C. Both A and B
    D. Neither A nor B

45. A vehicle pulls left after the brake pedal is released. This could be caused by:

    A. A restricted left front brake hose.
    B. A cut master cylinder piston seal.
    C. A faulty metering valve.
    D. A restricted right front brake hose.

PREPARATION EXAM 2

1. The brakes are dragging on a vehicle. This could be caused by:

    A. Overfilled master cylinder reservoir.
    B. Low vacuum to the booster.
    C. Bypassing master cylinder.
    D. Quick take-up valve stuck open.

2. The brake pads are worn at an angle on a vehicle with a sliding brake caliper system. Which of the following is the most likely cause?

    A. Worn caliper piston
    B. Worn caliper bushings
    C. Leaking caliper seal
    D. Leaking caliper boot

3. A diagnostic trouble code is set for an open wheel speed sensor on the left front sensor. The technician measures the resistance at the sensor and finds it to be within specification. Which of the following is the most likely cause for the wheel speed sensor code?

    A. Faulty ABS computer
    B. Open fuse
    C. Open wiring from the computer to the wheel speed sensor
    D. Shorted left front ABS modulator valve

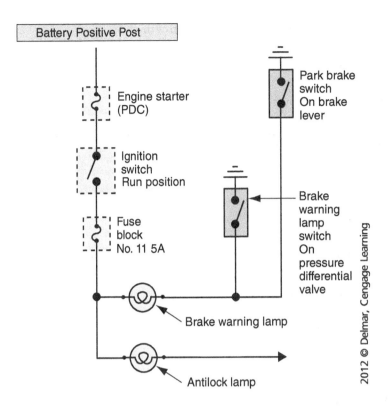

4. The brake warning lamp pictured above stays on any time the ignition switch is on. Technician A says this could be caused by a poor ground at the parking brake switch. Technician B says this could be caused by a poor ground at the brake warning lamp switch. Who is correct?

   A. A
   B. B
   C. Both A and B
   D. Neither A nor B

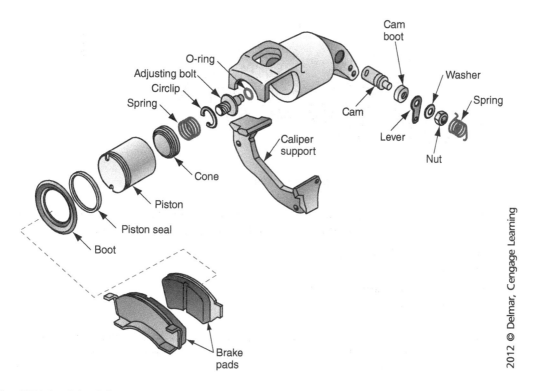

5. Which of the following is true concerning the above caliper?
   A. This caliper incorporates the parking brake.
   B. This caliper can only be used on the front.
   C. This caliper can only be used with Type 5 brake fluid.
   D. This caliper must be replaced at each brake pad replacement.

6. Technician A says a double flare line should be replaced with an ISO flare line. Technician B says an ISO flare line should be replaced with a double flare line. Who is correct?
   A. A
   B. B
   C. Both A and B
   D. Neither A nor B

Section 5 Sample Preparation Exams — Brakes (A5)

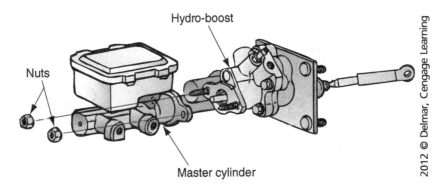

7. While testing the system shown above, the technician finds there is no assist from the booster after the engine is shut off. Which of the following is the most likely cause?

    A. A faulty back-up EH pump
    B. A faulty accumulator
    C. A bypassing master cylinder
    D. A bypassing brake booster

8. A customer is concerned that under heavy acceleration on dry pavement the traction control light comes on and the throttle momentarily becomes unresponsive. Which of the following is the most likely cause?

    A. A tire has lost traction.
    B. The accelerator pedal position (APP) sensor is faulty.
    C. The throttle position sensor (TPS) is faulty.
    D. A wheel speed sensor is faulty.

9. A vehicle has a low brake pedal. Master cylinder fluid level is correct. Which of the following is the most likely cause?

    A. Worn brake pad lining
    B. Seized wheel cylinder
    C. Seized caliper piston
    D. Brake shoes out of adjustment

10. The brake caliper piston boot was cut when the caliper was reassembled. Technician A says this will cause a brake fluid leak when the caliper is reinstalled. Technician B says this could cause the piston to bind due to contamination. Who is correct?

    A. A
    B. B
    C. Both A and B
    D. Neither A nor B

11. A parking brake cable is binding on a vehicle with front disc and rear drum brakes. This can result in:

    A. Dragging front brakes.
    B. Dragging rear brakes.
    C. Rusted brake drums.
    D. Rusted disc brake rotors.

12. Which of the following is LEAST LIKELY to be an input to the traction control system?
    A. Throttle position sensor
    B. Wheel speed sensor
    C. Engine RPM
    D. Transmission input shaft RPM

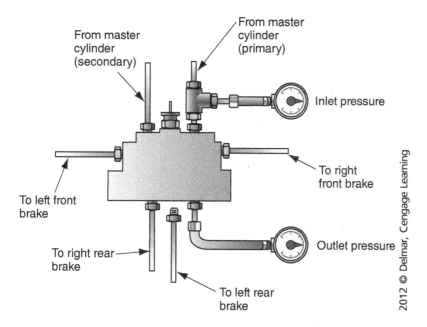

13. After a brake application, the inlet pressure gauge in the figure above drops to normal; however, the outlet pressure gauge remains higher than normal. Which of the following could be the cause?
    A. Binding brake pedal
    B. Incorrect brake pedal free stroke
    C. Restricted combination valve
    D. Weak brake shoe return springs

14. After disc brake pad replacement, the new pads should be burnished. This purpose of this procedure is to:
    A. Conform the pad lining to the rotor.
    B. Break the sealant off of the new linings.
    C. Remove finger prints from the rotor.
    D. Boil all the water out of the brake fluid.

15. Technician A says some wheel speed sensors are incorporated with the wheel bearing assembly and must be replaced as a unit. Technician B says some wheel speed sensors are located in the rear axle differential. Who is correct?
    A. A
    B. B
    C. Both A and B
    D. Neither A nor B

16. Which of the following is true concerning caliper removal?
    A. The fluid must be drained prior to removal.
    B. The caliper sliding pins must be removed prior to removal.
    C. The piston must be replaced every time the caliper is removed.
    D. The boot must be replaced every time the caliper is removed.

17. The brake fluid has been contaminated with oil. Which of the following is correct?
    A. All rubber components must be replaced.
    B. Only rubber components which show contamination should be replaced.
    C. All steel lines must be replaced.
    D. Only steel lines which show contamination should be replaced.

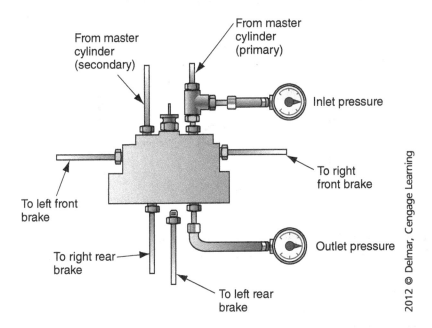

18. During a pressure test both gauges in the figure above read lower than normal. Technician A says a frozen (stuck) disc brake caliper could be the cause. Technician B says a frozen (stuck) wheel cylinder piston could be the cause. Who is correct?

    A. A
    B. B
    C. Both A and B
    D. Neither A nor B

19. When driving a new race of a tapered wheel bearing into the hub, which of the following is true?

    A. Drive the race in until flush.
    B. Drive the race in until a distinct sound change occurs.
    C. Drive the race in until it is just below the surface.
    D. Drive the race in using the old wheel bearing as a driving tool.

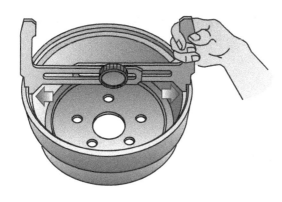

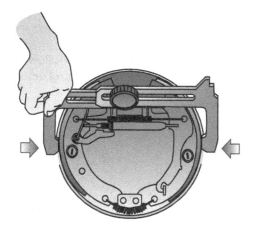

20. The tool shown above is used to:
    A. Arc the brake shoes.
    B. Compress the brake shoes.
    C. Compare the shoes to the drum.
    D. Refinish the brake drum.

21. When the brake line is loosened at the master cylinder no fluid will drip from the loosened line; however, when the brake pedal is pulled up fluid will start to drip. Which of the following is indicated?
    A. The master cylinder should be replaced.
    B. The master cylinder should be rebuilt.
    C. The brakes may be out of adjustment.
    D. The brake pedal may be out of adjustment.

22. When reinstalling the caliper slide pins, they should be lubricated with:
    A. High temperature wheel bearing grease.
    B. High temperature silicon grease.
    C. Lubriplate®
    D. Antiseize

Section 5 Sample Preparation Exams	Brakes (A5)

23. All of the following would cause the red brake warning lamp (RBWL) to be illuminated EXCEPT:
    A. A leaking wheel cylinder.
    B. A faulty master cylinder.
    C. Broken brake hose.
    D. Low pressure to the hydraulic brake booster.

24. Technician A says lug nuts for steel wheels should be torqued in a star pattern. Technician B says lug nuts for aluminum wheels should be torqued in a circle pattern. Who is correct?
    A. A
    B. B
    C. Both A and B
    D. Neither A nor B

25. Technician A says the purpose of the traction control deactivation switch is to allow wheel spin in deep snow or mud. Technician B says the purpose of the traction control deactivation switch is to improve fuel economy. Who is correct?

    A. A
    B. B
    C. Both A and B
    D. Neither A nor B

26. The brake fluid has been contaminated with oil. Technician A says all rubber hydraulic brake components must be replaced. Technician B says all the antilock braking system (ABS) wheel speed sensors must be replaced. Who is correct?
    A. A
    B. B
    C. Both A and B
    D. Neither A nor B

27. When adjusting the parking brake which of the following is correct?
    A. Adjust the cable first, then the shoes.
    B. Adjust the shoes first, then the cable.
    C. Adjust the calipers first, then the shoes.
    D. Adjust the calipers first, then the cable.

28. The traction control lamp stays on continuously on a vehicle. The ABS light is not lit. Which of the following is the most likely cause?

    A. Wheel speed sensor
    B. ABS control module
    C. Traction control module
    D. ABS modulator valve

29. Technician A says that a master cylinder pushrod that is adjusted too long could cause the brakes to drag. Technician B says that a master cylinder pushrod that is adjusted too long could result in repeat master cylinder failure. Who is right?

    A. A only
    B. B only
    C. Both A and B
    D. Neither A nor B

30. The caliper slide pin bushings are worn. Which of the following is true?
    A. The bushings must be replaced.
    B. The caliper must be replaced.
    C. The bushings and the caliper piston must be replaced.
    D. The bushings must be heated to be removed.

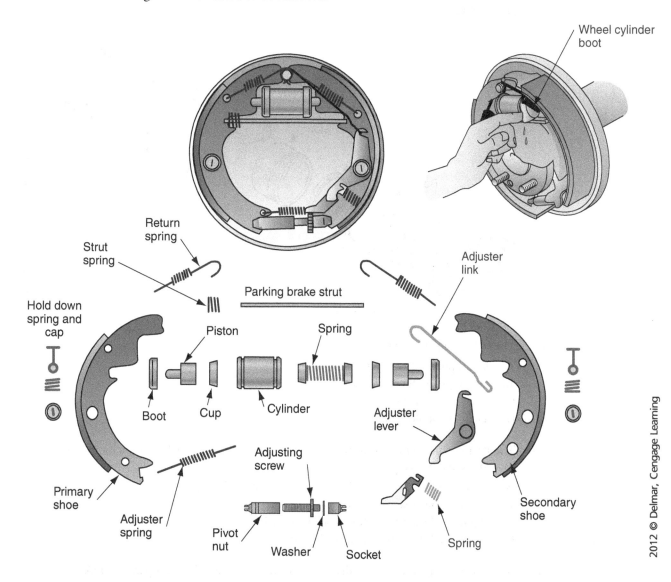

31. In the figure above, if the spring is left out during assembly, which of the following would most likely occur?
    A. The parking brake would not work.
    B. The brake would not self adjust.
    C. The service brake would only work in a rearward rotation.
    D. The service brake would only work in a forward rotation.

32. Vacuum to the booster is being measured with the engine idling. Which of the following would be considered a normal reading?

    A. 5" Hg
    B. 11" Hg
    C. 18" Hg
    D. 25" Hg

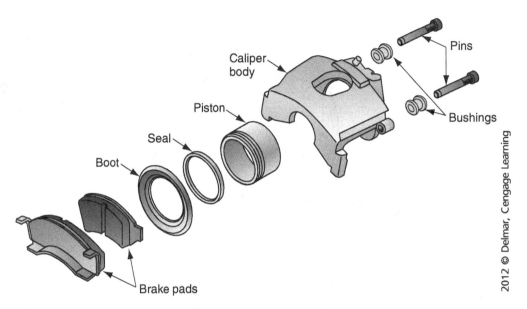

33. The brake caliper shown above is a:

    A. Quad piston floating design.
    B. Single piston floating design.
    C. Dual piston fixed design.
    D. Dual piston floating design.

34. The flexible line connected to the front disc brake caliper is cracked. Which of the following should the technician do?

    A. Repair the section of cracked hose using a barb fitting.
    B. Replace the hose with a new one.
    C. Replace all the rubber hoses on the vehicle.
    D. Replace all the rubber and steel hoses on the vehicle.

35. A vehicle equipped with front disc/rear drum brakes has a binding parking brake cable. This can result in:

    A. Difficulty removing the disc brake pads.
    B. Difficulty removing the brake drum.
    C. Difficulty installing the disc brake pads.
    D. Difficulty installing the brake shoes.

36. A customer is concerned that the traction control light flashed on momentarily while driving on slick roads. The technician finds no stored diagnostic trouble codes. Which of the following is most likely the cause?

    A. A faulty wheel speed sensor
    B. A faulty brake modulator
    C. Normal operation
    D. Mismatched tires

37. A step bore master cylinder has been diagnosed to be faulty. Technician A says that the brake booster must be replaced along with the master cylinder. Technician B says that the step bore master cylinder cannot be rebuilt because it cannot be honed. Who is right?

    A. A only
    B. B only
    C. Both A and B
    D. Neither A nor B

38. The hydraulic brakes on a car function correctly. However, the vacuum brake booster is not providing any brake assistance. Which of the following could be the cause?

    A. Too much vacuum to the booster
    B. Low brake fluid
    C. Incorrect brake fluid
    D. Too little vacuum to the booster

39. When installing the slide pins on the disc brake caliper the technician finds one hole stripped out. Which of the following is the most economical repair?

    A. The caliper must be replaced.
    B. The disc brake rotor must be replaced.
    C. There may be oversized bolts available.
    D. The steering knuckle must be replaced.

40. Technician A says DOT 3 and DOT 5 brake fluid can be mixed in a brake system without any problems. Technician B says DOT 4 and DOT 5 brake fluid can be mixed in a brake system without any problems. Who is correct?

    A. A
    B. B
    C. Both A and B
    D. Neither A nor B

41. The proper procedure for adjusting tapered roller wheel bearings used on some drum brake hubs is:

    A. Tighten while spinning the hub to seat the bearings, then back off to the specified amount of end-play.
    B. Finger tighten the adjusting nut.
    C. Torque the adjusting nut to 20 lb/ft of torque and then turn the nut an additional 45 degrees.
    D. Torque the nut to 75 ft lbs with the wheels on the ground.

42. A drum brake will rotate freely forward but will bind when rotated backward. Which of the following is the most likely cause?

    A. A broken return spring on the front shoe
    B. A broken return spring on the rear shoe
    C. A binding wheel bearing
    D. An out of round drum

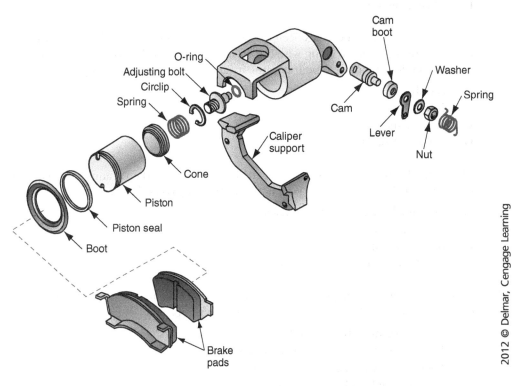

43. Refer to the illustration above. The service brakes work correctly; however, the parking brake will not function. Technician A says the problem could be a torn boot. Technician B says the problem could be air in the hydraulic system. Who is correct?

    A. A
    B. B
    C. Both A and B
    D. Neither A nor B

44. During a brake inspection on a single piston floating caliper the technician finds the inner pad worn much more than the outer. Technician A says a binding piston could be the cause. Technician B says a sticking caliper body could be the cause. Who is correct?

    A. A
    B. B
    C. Both A and B
    D. Neither A nor B

45. Brake fluid test strips can test the brake fluid for:

    A. Oil.
    B. Water.
    C. Freeze protection.
    D. Power steering fluid.

PREPARATION EXAM 3

1. The goal of burnishing the brakes after a complete brake service is to:
   A. Heat the brake assembly.
   B. Cool the brake assembly.
   C. Set the wheel bearings.
   D. Stretch the brake shoes to match the drum.

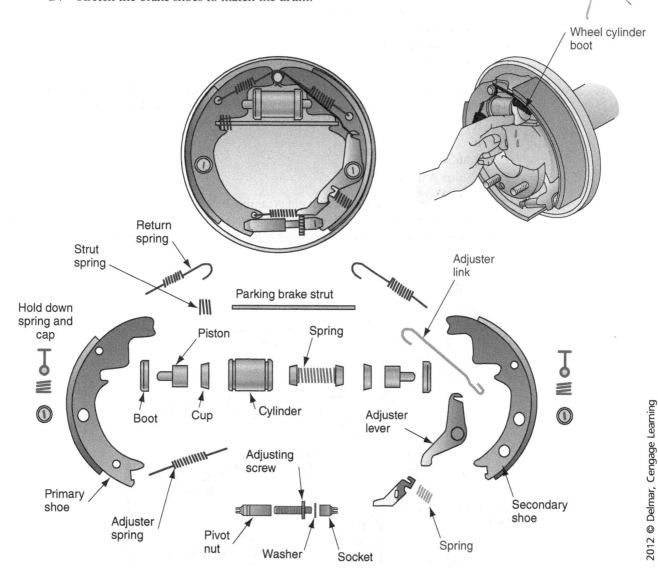

2. Refer to the figure above. The internal spring in the wheel cylinder is left out during a brake rebuild. Which of the following would be the most likely result?
   A. Air would be allowed in the brake fluid.
   B. The brake would drag.
   C. The brake would not apply.
   D. The brake would overheat.

3. Which of the following is true concerning master cylinder pushrod adjustment?
   A. There should be a slight gap between the pushrod and master cylinder piston.
   B. It should be adjusted until the piston is pushed in slightly.
   C. It should keep a slight pressure in the hydraulic system.
   D. It should be adjusted at each brake pad replacement.

4. Vacuum assist from the brake booster is present with the engine running but not present after the vehicle's engine is shut off. Which of the following could cause this condition?
   A. A faulty vacuum check valve
   B. A faulty vacuum booster hose
   C. Low manifold vacuum
   D. A leaking intake manifold

5. Technician A says an analog wheel speed sensor generates a square wave output signal. Technician B says a digital wheel speed sensor generates an ac signal. Who is correct?
   A. A
   B. B
   C. Both A and B
   D. Neither A nor B

6. Technician A says if a vehicle has a floating rear disc brake caliper the parking brake actuation mechanism will be inside the caliper piston. Technician B says if a vehicle has a fixed rear disc brake caliper the parking brake may be inside the disc brake rotor. Who is correct?
   A. A
   B. B
   C. Both A and B
   D. Neither A nor B

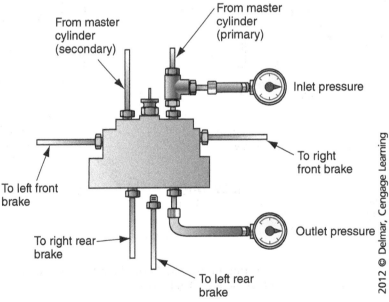

7. After a brake application, the inlet pressure gauge in the figure above drops to normal; however, the outlet pressure gauge remains higher than normal. Technician A says a faulty disc brake caliper could be the cause. Technician B says a faulty wheel cylinder could be the cause. Who is correct?
   A. A
   B. B
   C. Both A and B
   D. Neither A nor B

8. The minimum rotor thickness specification is 1.250" (31.75mm). Which of the following measurements would indicate the rotor is reusable?

    A. 1.115" (28.32mm)
    B. 1.125" (28.57mm)
    C. 1.225" (31.11mm)
    D. 1.520" (38.61mm)

9. When the parking brake is released the parking brake pedal fails to return to its most upward position. Technician A says a faulty master cylinder could be the cause. Technician B says air trapped in the brake fluid could be the cause. Who is correct?

    A. A
    B. B
    C. Both A and B
    D. Neither A nor B

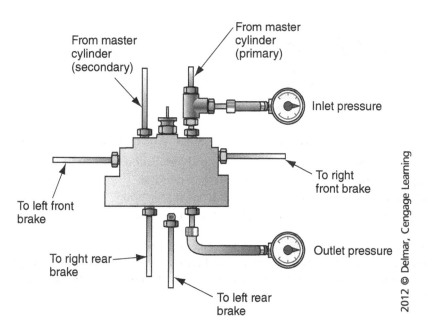

10. A vehicle with poor brake performance is being tested as shown above. Both pressure gauges are below normal. Which of the following could be the cause?

    A. A restricted combination valve
    B. A leaking cup on the primary piston
    C. A leaking cup on the secondary piston
    D. A restricted pressure differential switch

11. A complete brake job including installing a new master cylinder has been performed on a vehicle. After a short test drive all the brakes locked up. Which of the following is the most likely cause?

    A. Brake shoes were incorrectly installed.
    B. Brake pads were incorrectly installed.
    C. The master cylinder was incorrectly installed.
    D. The antilock braking system (ABS) sensors were not properly reset.

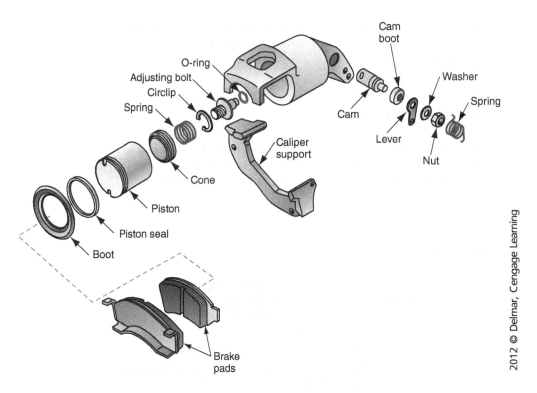

12. The brake pads need to be replaced on the above system. Which of the following is true?

    A. The adjusting bolt must be retracted prior to pad installation.
    B. The adjusting bolt must be retracted prior to pad removal.
    C. The adjusting bolt must be extended prior to pad installation.
    D. The adjusting bolt must be extended prior to pad removal.

13. When a customer attempts to apply the parking brake, the pedal travels its full range of motion however the parking brakes do not apply. Technician A says a faulty master cylinder could be the cause. Technician B says air trapped in the brake fluid could be the cause. Who is correct?

    A. A
    B. B
    C. Both A and B
    D. Neither A nor B

14. Which of the following is true concerning a non-integral ABS system?

    A. The master cylinder and booster are mounted separately.
    B. The ABS controller and master cylinder are mounted together.
    C. They can use a vacuum brake booster.
    D. They use an electro-hydraulic pump as a booster.

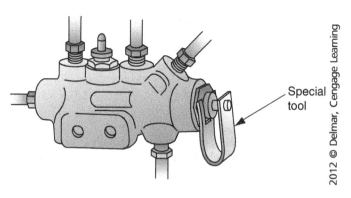

15. The tool shown above is used on the:
    A. Master cylinder.
    B. Brake booster.
    C. Combination valve.
    D. Proportioning valve.

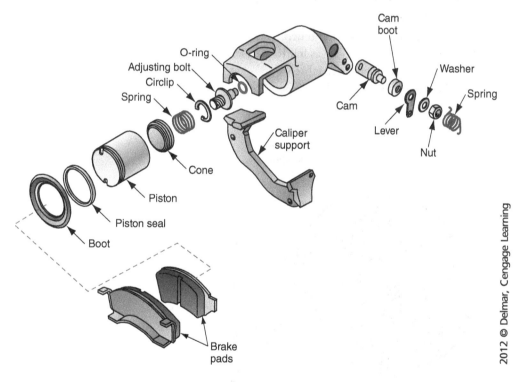

16. Refer to the figure above. Which of the following would normally be used to return the piston fully into the caliper bore?
    A. A flat tip screwdriver
    B. A cross tip (Phillips) screw driver
    C. A special tool
    D. A C-clamp

17. The screw threads of the drum brake self adjuster should be lubricated with which of the following?

    A. Oil
    B. Chassis grease
    C. Lubriplate®
    D. Antiseize

18. The right front caliper is inoperative. All other brakes operate normally. Technician A says that a collapsed brake hose could cause this. Technician B says that a faulty master cylinder could cause this. Who is correct?

    A. A only
    B. B only
    C. Both A and B
    D. Neither A nor B

19. Tapered front roller bearings were replaced during a brake job. The vehicle has come back into the shop with the wheel bearing welded to the spindle. Which of the following is the most likely cause?

    A. The bearing was overtightened.
    B. The bearing clearance was left too loose.
    C. The lug nuts were overtightened.
    D. The lug nuts were left loose.

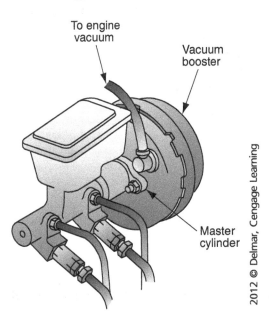

20. Technician A says the line labeled "To engine vacuum" in the figure above would connect to the intake manifold on a diesel engine. Technician B says the line labeled "To engine vacuum" would connect to the intake manifold on a gasoline engine. Who is correct?

    A. A
    B. B
    C. Both A and B
    D. Neither A nor B

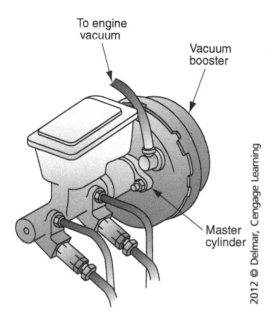

21. In the picture shown above, the two lower brake lines are equipped with:

    A. Proportioning valves.
    B. Metering valves.
    C. Pressure differential valves.
    D. Antilock braking system (ABS) valves.

22. Which of the following is most likely to use a single channel ABS system?

    A. Front-wheel drive
    B. Rear-wheel drive
    C. Four-wheel drive
    D. All-wheel drive

23. A tapered wheel bearing has failed and spun on the spindle. The spindle now is blue/purple where the inner race sits. Besides replacing the bearing what else must be replaced?

    A. The brake pads
    B. The rotor
    C. The hub
    D. The spindle

24. On a vehicle with rear disc brakes using an integral parking brake, the service brakes work correctly; however, the parking brake will not function. Technician A says the problem could be a seized caliper piston. Technician B says the problem could be a leaking piston seal. Who is correct?

    A. A
    B. B
    C. Both A and B
    D. Neither A nor B

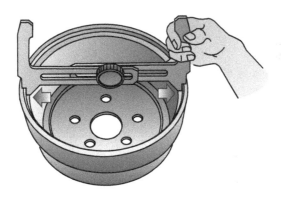

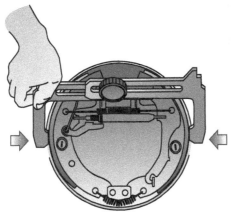

25. When using the tool shown above a technician discovers that the shoes are larger than the drum. Which of the following should the technician do next?

   A. Replace the shoes.
   B. Replace the drum.
   C. Turn the drum.
   D. Turn the adjuster.

26. The technician measures the front rotors; the results are listed below:

   Minimum Rotor Thickness, Specification: .898″ (22.80mm)
   Actual Rotor Thickness, Left: .988″ (25.09mm)
   Actual Rotor Thickness, Right: .989″ (25.12mm)

   Which of the following should be done?

   A. Replace the left rotor
   B. Replace both rotors
   C. Reuse both rotors
   D. Replace the right rotor

27. Technician A says some parking brakes have a pedal to push to apply the parking brake. Technician B says some parking brakes have a pedal to push to release the parking brake. Who is correct?

    A. A
    B. B
    C. Both A and B
    D. Neither A nor B

28. A single channel ABS system will have ABS installed on:

    A. The front wheels only.
    B. The left side only.
    C. The rear wheels only.
    D. The right side only.

29. The left front wheel bearing has been replaced on a front wheel drive vehicle. Which of the following is true concerning the correct procedure to tighten the drive axle nut?

    A. The nut should be tightened then backed off one flat.
    B. The nut should be tightened to the torque specification.
    C. The nut should be adjusted to allow the wheel bearing to have 0.0010″–0.0050″ (0.0254mm–0.127mm) end-play.
    D. The self locking nut should be reused.

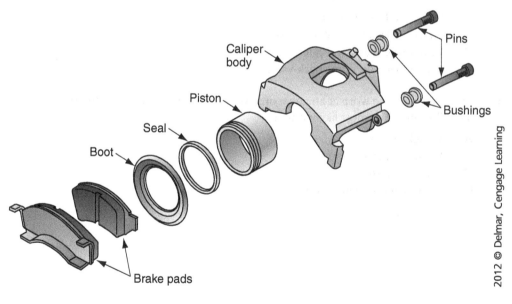

30. Technician A says the above pads can be removed without removing the caliper. Technician B says the above pads can be removed without removing the pins. Who is correct?

    A. A
    B. B
    C. Both A and B
    D. Neither A nor B

31. Technician A says that a leaking wheel cylinder can be identified by looking behind the boot. Technician B says that some wheel cylinders cannot be honed. Who is correct?

    A. A only
    B. B only
    C. Both A and B
    D. Neither A nor B

32. Which of the following is true concerning an integral ABS system?

    A. The master cylinder and booster are separately mounted.
    B. The ABS controller and master cylinder are separately mounted.
    C. They use a vacuum brake booster.
    D. They use an electro-hydraulic pump.

33. The left front disc brake caliper is dragging. All the other brakes operate normally. Which of the following could be the cause?

    A. Damaged right front brake line
    B. Damaged left front brake line
    C. Faulty master cylinder primary piston seal
    D. Faulty master cylinder secondary piston seal

34. The brake pedal on a vehicle will slowly drop while the vehicle is sitting at a stop light. Which of the following could be the cause?

    A. A restricted vacuum hose
    B. Externally leaking master cylinder
    C. A bypassing master cylinder
    D. A stuck open vacuum check valve

35. The brake pads are worn at an angle on a vehicle equipped with rear disc brakes. Which of the following is the most likely cause?

    A. A stuck piston
    B. A bent caliper support
    C. A stuck adjuster
    D. A bent parking brake lever

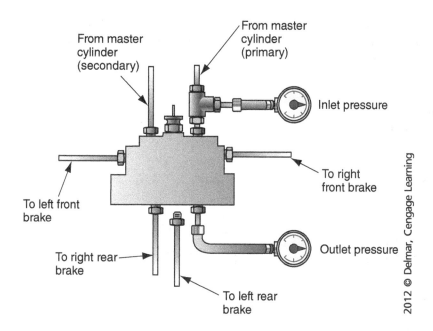

36. During a pressure test both gauges in the figure above read lower than normal. Technician A says a swollen hose could be the cause. Technician B says a restricted hose could be the cause. Who is correct?

   A. A
   B. B
   C. Both A and B
   D. Neither A nor B

37. Technician A says an analog wheel speed sensor can be checked with a DMM set on AC volts. Technician B says an analog wheel speed sensor can be checked with an ohmmeter. Who is correct?

   A. A
   B. B
   C. Both A and B
   D. Neither A nor B

38. A hydro-boost system is being checked on a vehicle. The power steering system functions normally; however, the brakes have very little power assist. Which of the following could be the cause?

   A. Faulty hydro-boost unit
   B. Faulty power steering pump
   C. Leaking steering gear
   D. Sticking tensioner

39. When installing threaded caliper slide pins, which of the following is the most normal tightening procedure?

   A. Tighten to 100 ft lbs, then tighten to 250 ft lbs.
   B. Tighten to 100 ft lbs then back off 1/2 turn.
   C. Tighten to 25 ft lbs.
   D. Tighten to 50 ft lbs and back off one turn.

40. The brake pedal of a vehicle slowly falls to the floor. No leaks are found and the reservoir is full. Technician A says that the cup seals in the master cylinder may be bypassing. Technician B says that a corroded internal master cylinder bore could be the cause. Who is correct?

    A. A only
    B. B only
    C. Both A and B
    D. Neither A nor B

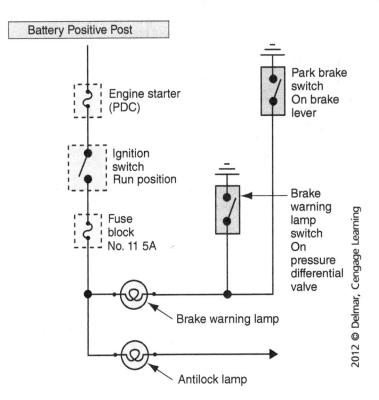

41. The brake warning lamp pictured above will come on when the service brakes are applied. Which of the following could be the cause?

    A. A misadjusted parking brake switch
    B. Air in the brake fluid
    C. Worn parking brakes
    D. Worn brake drums

42. All of the following could cause the wheel cylinder to leak EXCEPT:

    A. Brake pedal adjustment.
    B. Lack of fluid maintenance.
    C. Torn dust boots.
    D. Recent shoe replacement.

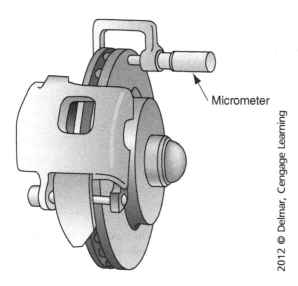
Micrometer

43. Technician A says rotor thickness is being measured in the figure. Technician B says rotor runout is being measured. Who is correct?

   A. A
   B. B
   C. Both A and B
   D. Neither A nor B

44. A brake system which uses a vacuum brake booster has poor braking performance and a hard, high brake pedal with the engine running. Which of the following could be the cause?

   A. A restricted vacuum hose
   B. A leaking master cylinder
   C. A bypassing master cylinder
   D. A stuck open vacuum check valve.

45. When bench bleeding the master cylinder which statement is true?

   A. The piston should be stroked until bubbles start to flow.
   B. The piston should be stroked until bubbles stop flowing.
   C. Mineral spirits should be used.
   D. Mineral oil should be used.

PREPARATION EXAM 4

1. Four channel ABS sensors will have how many ABS wheel speed sensors?

   A. Two
   B. Four
   C. Six
   D. Eight

2. The metering valve:
   A. Limits front brake application until the rear brakes apply.
   B. Limits rear brake application until the front brakes apply.
   C. Limits front brake application based on vehicle height.
   D. Limits rear brake application based on vehicle height.

3. A wheel cylinder has been honed. It should be cleaned with:
   A. Alcohol.
   B. Mineral spirits.
   C. Penetrating solvent.
   D. Rust preventative.

4. A vehicle is equipped with a disc/drum brake system. When the customer pulls the parking brake lever the lever travels its full range of motion, but the parking brake does not apply. Technician A says the disc brakes could be out of adjustment. Technician B says the drum brakes could be out of adjustment. Who is correct?
   A. A
   B. B
   C. Both A and B
   D. Neither A nor B

5. A height sensing proportioning valve would be located on:
   A. The left front.
   B. The right front.
   C. The rear in the center.
   D. Only on all wheel drive vehicles.

6. The customer feels a rapid pedal pulsation under normal braking conditions. This could be caused by all of the following EXCEPT:
   A. Different size tires.
   B. A damaged reluctor wheel.
   C. Excessive reluctor to sensor gap.
   D. Open circuit to the speed sensor.

7. The technician measures the front rotors; the results are listed below:

   Minimum Rotor Thickness, Specification: .898" (22.80mm)
   Actual Rotor Thickness, Left: .889" (22.58mm)
   Actual Rotor Thickness, Right: .809" (25.12mm)

   Which of the following should be done?
   A. Replace the left rotor.
   B. Replace both rotors.
   C. Reuse both rotors.
   D. Replace the right rotor.

8. Technician A says the primary shoe goes toward the front of the vehicle. Technician B says the secondary shoe goes toward the rear of the vehicle. Who is right?

   A. A
   B. B
   C. Both A and B
   D. Neither A nor B

9. Which of the following would most commonly be used as the drive axle retaining nut on a front wheel drive vehicle?

   A. Self-locking nylon insert nut
   B. Plain hex nut without a locking feature
   C. Double nuts with a lock between them
   D. Self-locking crimp style nut

10. The rotor thickness specification is 1.250" (31.75mm). In three places the rotor measures 1.350" (34.29mm). In a fourth place the rotor measures 1.248" (31.70mm). Which of the following is true?

    A. The rotor should be turned to the smallest measurement and reused.
    B. The rotor should be reused as is.
    C. The rotor should be replaced.
    D. The measurement was performed incorrectly.

11. Technician A says the same tool is used to make a double flare or an ISO flare. Technician B says the same tool is used to make a double flare and a single flare. Who is correct?

    A. A
    B. B
    C. Both A and B
    D. Neither A nor B

12. Which of following statements is true concerning the brake warning lights?

    A. The master cylinder can cause the amber light to illuminate.
    B. The master cylinder can cause the red light to illuminate.
    C. If the amber light is illuminated the vehicle must be pulled to the side of the road and towed.
    D. If the amber light is illuminated there is a possible hydraulic imbalance.

13. The technician is preparing to do a brake job on a vehicle equipped with a non-integral ABS system. Which of the following should the technician do?

    A. Pump the pedal 25–40 times.
    B. Support the vehicle on a jack.
    C. Support the vehicle on jack stands.
    D. Release the vacuum from the brake booster.

14. The technician has measured the rotor in 12 locations. Which variation in the readings would indicate a failed rotor?

    A. 0.0001" (0.00254mm)
    B. 0.0005" (0.0127mm)
    C. 0.001" (0.0254mm)
    D. 0.005" (0.127mm)

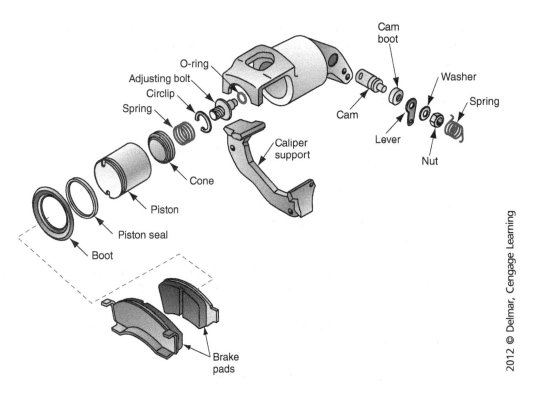

15. In the figure above, the lever would be connected to:
    A. The master cylinder.
    B. The sway bar.
    C. The parking brake cable.
    D. The brake pedal.

16. The flexible rubber brake hose connecting to the caliper is being replaced. Which of the following should also be replaced?
    A. The caliper
    B. The banjo bolt
    C. The sealing washers
    D. The steel brake line

17. An ABS hydraulic control unit (HCU) makes a clicking noise for approximately 5 seconds after starting the vehicle. Which of the following is indicated?
    A. Faulty ABS module.
    B. Faulty ABS hydraulic unit.
    C. The ABS electronic control module (ECM) is bleeding the brakes.
    D. The ABS ECM is performing a self-test.

18. Technician A says the primary shoe has the long pad. Technician B says the secondary shoe is the short pad. Who is correct?
    A. A
    B. B
    C. Both A and B
    D. Neither A nor B

19. Which is true concerning measuring a rotor for thickness variation?
    A. The rotor should be hot when this measurement is taken.
    B. The wheel bearing should be tightened when the measurement is taken.
    C. The measurement should be taken in several locations around the rotor.
    D. The caliper must be removed before this measurement can be taken.

20. The master cylinder cover is removed and the steering wheel is rocked back and forth. The fluid in the master cylinder moves as the steering wheel is turned. Which of the following is the most likely cause?
    A. Warped brake rotor
    B. Worn caliper piston
    C. Seized sliding caliper
    D. Loose wheel bearings

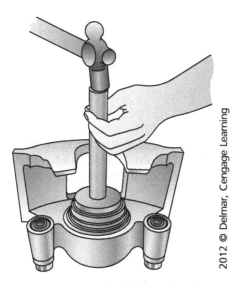

21. What is being installed in the figure above?
    A. The piston
    B. The piston seal
    C. The boot
    D. The square cut o-ring

22. Technician A says ABS trouble codes can usually be retrieved using a scan tool. Technician B says ABS codes can usually be cleared using a scan tool. Who is correct?
    A. A
    B. B
    C. Both A and B
    D. Neither A nor B

23. The right rear brake is dragging on a vehicle with independent rear suspension. All other brakes operate normally. Which of the following is could be the cause?
    A. Damaged right front brake line
    B. Damaged right rear brake line
    C. Damaged left rear brake line
    D. Damaged left front brake line

24. The technician is counting the "clicks" before the parking brake applies. Which of the following is generally considered the maximum number for a proper parking brake adjustment?
    A. 5
    B. 10
    C. 15
    D. 20

25. What is the proper brake bleeding sequence?
    A. RR, LF, LR, RF
    B. LR, RF, LF, RR
    C. LR, LF, RR, RF
    D. RR, LR, RF, LF

26. A pressure gauge is installed in the bleeder screw port of a disc brake caliper, the brake pedal is depressed, and the gauge pressure indicates 750 psi. The pedal is held steady and the engine is started; the gauge pressure does not change. Which of the following is indicated?

    A. This is a normal condition.
    B. The combination valve is leaking.
    C. The combination valve is restricted.
    D. The brake booster is not functioning.

27. Which of the following would best describe the burnishing procedure?
    A. Five complete stops from 30 MPH
    B. Twenty slow-downs from 50 MPH to 30 MPH
    C. Five slow-downs from 50 MPH to 30 MPH
    D. Fifty slow-downs from 50 MPH to 30 MPH

28. With the vehicle supported on a frame hoist and the ignition on, the technician spins one front wheel by hand. The ABS lamp illuminates. This indicates:
    A. A faulty front wheel speed sensor.
    B. A faulty rear wheel speed sensor.
    C. A faulty ABS control unit.
    D. Normal operation.

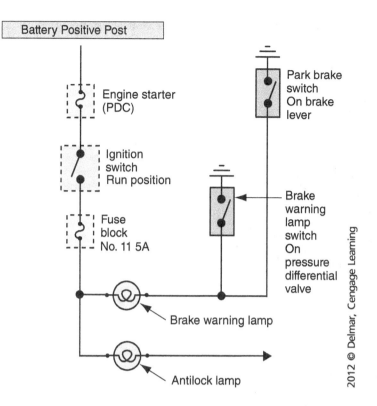

29. The brake warning lamp pictured above stays on any time the ignition switch is on. Technician A says this could be caused by an open at the parking brake switch. Technician B says this could be caused by an open at the brake warning lamp switch. Who is correct?

    A. A
    B. B
    C. Both A and B
    D. Neither A nor B

30. Technician A says after bench bleeding the master cylinder it can be reinstalled. Technician B says after bench bleeding, the master cylinder should be emptied. Who is correct?

    A. A
    B. B
    C. Both A and B
    D. Neither A nor B

31. While testing a vehicle equipped with a vacuum brake booster, if the engine is shut off and pressure applied to the brake pedal the pedal is hard and firm. Otherwise the brakes function normally. Which of the following is indicated by this test result?

    A. A restricted vacuum hose
    B. A leaking master cylinder
    C. A leaking vacuum check valve
    D. Air in the brake system

32. A vehicle had worn pads and they were replaced; now the vehicle's stopping power is poor. This is most likely caused by:

    A. Contaminated brake fluid.
    B. Air in the system.
    C. Improper break-in procedure.
    D. Faulty master cylinder.

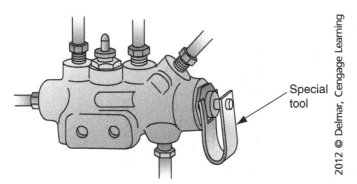

Special tool

33. The tool shown above is most likely used when:

    A. Bleeding the brakes.
    B. Bleeding the brake booster.
    C. Diagnosing a dash brake light concern.
    D. Diagnosing an ABS system concern.

34. Which one of the following methods of brake bleeding typically requires two technicians?

    A. Gravity bleeding
    B. Brake pedal bleeding
    C. Vacuum bleeding
    D. Pressure bleeding

35. The hydro-boost belt driven pump is being tested. Which of the following would normally be considered the maximum pump pressure specification?

    A. 750 psi
    B. 1000 psi
    C. 1500 psi
    D. 2000 psi

36. Which of the following would most commonly be used as the spindle nut on a set of tapered roller bearings?

    A. Self-locking nylon insert nut
    B. Castellated (slotted) nut with a cotter pin
    C. Double nuts with a lock between them
    D. Self-locking crimp style nut

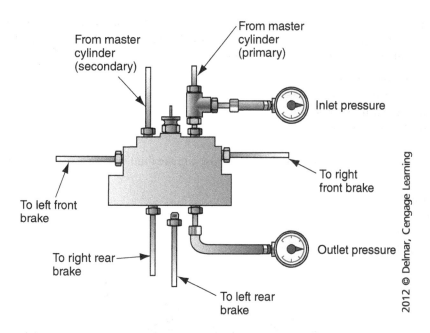

37. After a brake application, the inlet pressure gauge in the figure above drops to normal; however, the outlet pressure gauge remains higher than normal. Technician A says a faulty master cylinder could be the cause. Technician B says a faulty parking brake could be the cause. Who is correct?

    A. A
    B. B
    C. Both A and B
    D. Neither A nor B

38. Brake shoe lubricant should be applied:

    A. To the shoe where it contacts the drum.
    B. To the shoe where it contacts the backing plate.
    C. To the drum where it contacts the hub.
    D. To the drum where it contacts the axle.

39. A vehicle equipped with a hydro-boost system is leaking fluid where the booster and master cylinder meet. Which of the following is true?

    A. A leak here can only come from the booster assembly.
    B. A leak here can only come from the master cylinder.
    C. A pink fluid here would indicate a brake fluid leak.
    D. A pink fluid here would indicate a power steering fluid leak.

40. What measurement is performed on a disc brake rotor using a single dial indicator?

    A. Rotor lateral runout
    B. Rotor radial runout
    C. Rotor thickness variation
    D. Rotor imbalance

41. While testing a brake system with a vacuum brake booster the technician finds that if the engine is started with pressure applied to the brake pedal the pedal will drop about 1/2 inch. Which of the following is indicated by this test result?

    A. A restricted vacuum hose
    B. A leaking master cylinder
    C. A bypassing master cylinder
    D. A normally operating booster

42. Which of the following would be most likely to cause a brake fluid leak on a disc brake caliper assembly?

    A. Piston seal
    B. Piston boot
    C. Caliper slide pins
    D. Caliper slide bushings

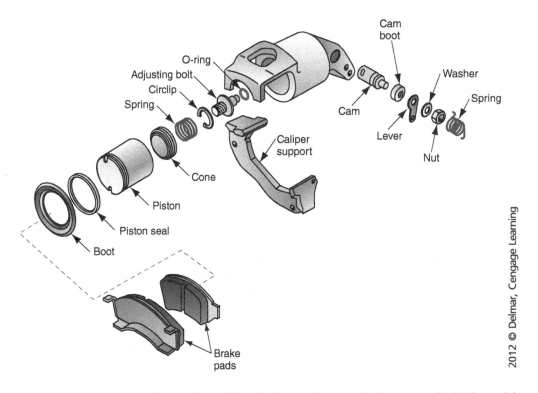

43. Refer to the illustration above. The parking brake works correctly; however, the brake pedal is soft and spongy when the service brake pedal is pushed. Technician A says the problem could be a torn boot. Technician B says the problem could be air in the hydraulic system. Who is correct?

    A. A
    B. B
    C. Both A and B
    D. Neither A nor B

44. When installing a disc brake caliper boot the technician should stop driving when:
    A. The driving tool is flush with the caliper.
    B. The driving tool is below the caliper surface.
    C. The seal is flush with the caliper.
    D. The sound changes to a dull thud.

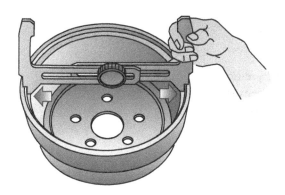

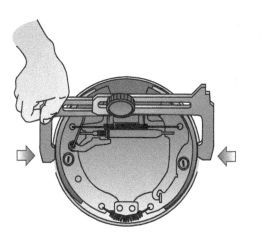

45. Technician A says the above tool is used to install the springs on the shoes. Technician B says the tool shown above is used to measure the wear lip inside the drum. Who is correct?
    A. A
    B. B
    C. Both A and B
    D. Neither A nor B

PREPARATION EXAM 5

1. Rotor lateral runout is greater than specification. This would most likely result in:
    A. Vibration while braking.
    B. Spongy brakes.
    C. A hard brake pedal.
    D. A low brake pedal.

2. Technician A says the parking brake adjustment may prevent the drum from fitting over new brake shoes. Technician B says some parking brakes are adjustable at the parking brake lever. Who is correct?

   A. A
   B. B
   C. Both A and B
   D. Neither A nor B

3. On a three-channel ABS system:

   A. The front wheels will share an ABS sensor.
   B. The rear wheels will share an ABS sensor.
   C. The left side will share an ABS sensor.
   D. The right side will share an ABS sensor.

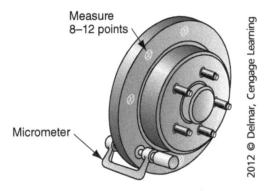

4. The micrometer above is measuring:

   A. Thickness variation.
   B. Lateral runout.
   C. Radial runout.
   D. Hardness.

5. The drum brake squeaks when the brakes are applied. Technician A says the shoes may be worn out. Technician B says the drum may be out of round. Who is correct?

   A. A
   B. B
   C. Both A and B
   D. Neither A nor B

6. A car with a vacuum brake booster is idling too fast. When the vacuum line going to the brake booster is crimped shut the idle decreases to normal. Which of the following is the most likely cause?

   A. The vacuum check valve is leaking.
   B. The metering valve is leaking.
   C. The proportioning valve is leaking.
   D. The brake booster is leaking.

7. A vehicle pulls to the left while braking. All of the following could cause this EXCEPT:
   A. Sticking right front caliper piston.
   B. Sticking metering valve.
   C. Worn suspension components.
   D. Restricted right front brake line.

8. The parking brake will not fully release. All of the following could be the cause EXCEPT:
   A. Air in the brake fluid.
   B. Binding parking brake cable.
   C. Broken return spring.
   D. Rusted actuator piston.

9. The disc brake rotor thickness variation is greater than specification. This would most likely result in:
   A. Pedal pulsation.
   B. Spongy brakes.
   C. A hard brake pedal.
   D. A low brake pedal.

10. The height sensing proportioning valve:
    A. Limits the rear brake pressure when the rear is fully loaded.
    B. Limits the rear brake pressure when the rear is lightly loaded.
    C. Limits the front brakes when the front is fully loaded.
    D. Limits the front brakes when the front is lightly loaded.

11. A vehicle is equipped with a vacuum brake booster. When the engine is running and the brakes are applied there is a constant sound of air movement (vacuum) around the brake pedal area. Technician A says the vacuum check valve is leaking. Technician B says the vacuum booster is leaking. Who is correct?
    A. A
    B. B
    C. Both A and B
    D. Neither A nor B

12. Technician A says some wheel speed sensors are adjusted using a feeler gauge. Technician B says some ABS wheel speed sensors are adjusted using a paper spacer, which comes with the new wheel speed sensor. Who is correct?
    A. A
    B. B
    C. Both A and B
    D. Neither A nor B

13. The technician is replacing a section of steel brake line. The brake line should be cut with a:
    A. Hacksaw.
    B. Tubing cutter.
    C. Torch.
    D. Pipe cutter.

14. Which of the following would be considered maximum thickness variation for a disc brake rotor?

    A. 0.0005"
    B. 0.002"
    C. 0.010"
    D. 0.025"

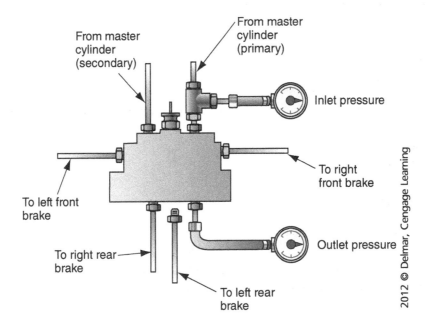

15. The gauge pressures read normal during the initial test. After several high speed, high-pressure brake applications the gauges read lower and began to fluctuate. Which of the following could be the cause?

    A. The brake fluid is contaminated with oil.
    B. The brake fluid is contaminated with water.
    C. The master cylinder is faulty.
    D. The wheel cylinders are leaking.

16. Which of the following would cause a glazed drum?

    A. A dragging shoe
    B. Air in the hydraulic system
    C. Missing brake shoe lining
    D. A frozen wheel cylinder

17. The parking brake will not apply. All of the following could be the cause EXCEPT:

    A. A broken parking brake cable.
    B. A broken return spring.
    C. A loose drum brake adjustment.
    D. A stretched parking brake cable.

18. Replacement metal brake lines should be constructed of:
    A. Copper.
    B. Steel.
    C. Aluminum.
    D. Brass.

19. Technician A says digital wheel speed sensors will have voltage supplied to them through the ABS control module. Technician B says digital wheel speed sensors are tested using an ohmmeter. Who is correct?
    A. A
    B. B
    C. Both A and B
    D. Neither A nor B

20. While installing the disc brake caliper boot it is ripped. Technician A says this will cause a brake fluid leak. Technician B says this will cause a low, soft brake pedal. Who is correct?
    A. A
    B. B
    C. Both A and B
    D. Neither A nor B

21. A vehicle which has poor brake performance also has a high, hard brake pedal. Technician A says there may be air in the system. Technician B says the brakes may be out of adjustment. Who is correct?
    A. A
    B. B
    C. Both A and B
    D. Neither A nor B

22. Technician A says installing a tire with a different diameter than what was originally installed on the vehicle may set an ABS code. Technician B says using DOT 3 instead of DOT 4 brake fluid may set an ABS code. Who is correct?
    A. A
    B. B
    C. Both A and B
    D. Neither A nor B

23. When the brake pedal is depressed, the rubber brake line connecting to the disc brake caliper swells. Which of the following is true?
    A. The caliper piston is stuck.
    B. The sliding caliper is stuck.
    C. The master cylinder is over pressurizing the hose.
    D. The hose should be replaced.

24. Technician A says wheel bearings which are adjusted too tight can cause a wheel to lockup. Technician B says wheel bearings which are adjusted too tight will set an antilock braking system (ABS) trouble code. Who is correct?
    A. A
    B. B
    C. Both A and B
    D. Neither A nor B

25. Which of the following could cause low vacuum to the booster with the engine running?

    A. A stuck open vacuum check valve
    B. A loose exhaust manifold
    C. A cracked intake manifold
    D. A stuck closed metering valve

26. When installing a disc brake caliper boot which of the following is true concerning tool lubrication?

    A. The tool should be coated with brake fluid.
    B. The tool should be coated with engine oil.
    C. The tool should be dry.
    D. The tool should be coated with lubricating oil.

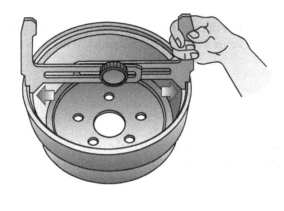

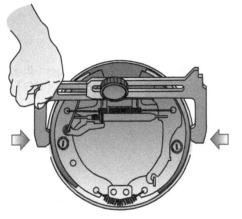

27. Technician A says the tool shown in the figure above is used to select the correct drum. Technician B says this tool is used to select the correct brake shoes. Who is correct?

    A. A
    B. B
    C. Both A and B
    D. Neither A nor B

28. After several brake applications all the brakes drag on a vehicle. Technician A says the compensating port may be covered. Technician B says the replenishing port may be covered. Who is correct?

    A. A
    B. B
    C. Both A and B
    D. Neither A nor B

29. A digital wheel speed sensor is being checked. When the sensor is disconnected the technician finds 5 V DC on the control module side of the connector. Which of the following is true?

    A. The voltage should be 12 V DC during this check.
    B. There should be no voltage present.
    C. There should be 2 V AC during this check.
    D. The test is being performed incorrectly.

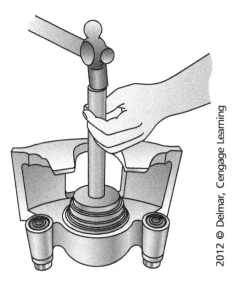

30. What is the technicain doing in the above illustration?

    A. Installing the boot
    B. Removing the piston
    C. Installing the piston
    D. Removing the boot

31. The parking brake will not apply. Which of the following could be the cause?

    A. Misadjusted parking brake cable
    B. Misadjusted stop light switch
    C. Leaking vacuum booster check valve
    D. Leaking vacuum booster

32. The purpose of bleeding the hydraulic system after a repair is:

    A. To remove moisture from the system.
    B. To flush old brake fluid from the system.
    C. To clean debris from the master cylinder.
    D. To remove air from the system.

33. The technician measures the rear rotors; the results are listed below:

    Minimum Rotor Thickness, Specification: 1.200″ (30.48mm)
    Actual Rotor Thickness, Left: 1.054″ (26.77mm)
    Actual Rotor Thickness, Right: 1.230″ (31.24mm)

    Which of the following should be done?

    A. Replace the left rotor.
    B. Replace both rotors.
    C. Reuse both rotors.
    D. Replace the right rotor.

34. When bleeding the brakes using the brake pedal, which of the following is true?

    A. Loosen the bleeder screw and pump the pedal.
    B. Make quick pedal applications then loosen the bleeder screw.
    C. Slowly apply the brake pedal, closing the bleeder screw before the end of the stroke.
    D. Stroke the pedal all the way to the floor.

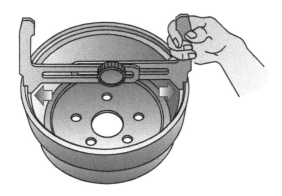

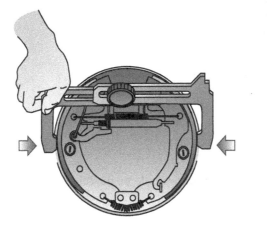

35. After the tool shown in the figure above is sized to the drum, it will not fit over the shoes. What should the technician do?

    A. Replace the shoes.
    B. Turn the drum.
    C. Replace the springs.
    D. Adjust the shoes.

36. Technician A says the parking brakes should be adjusted before the service brakes. Technician B says that a stretched parking brake cable can cause the parking brake to drag. Who is correct?

    A. A
    B. B
    C. Both A and B
    D. Neither A nor B

37. Which of the following would indicate maximum allowable runout of a disc brake rotor?

    A. 0.0005″
    B. 0.0002″
    C. 0.0004″
    D. 0.0020″

38. Technician A says a scan tool may be necessary to bleed the brakes on an ABS equipped vehicle. Technician B says bleeding brakes on an ABS equipped vehicle may be a two-technician job. Who is correct?

    A. A
    B. B
    C. Both A and B
    D. Neither A nor B

39. When the brake pedal is depressed fluid leaks from the rear of the master cylinder. This could be caused by:

    A. A leaking primary piston seal.
    B. A leaking secondary piston seal.
    C. A leaking primary cup.
    D. A leaking secondary cup.

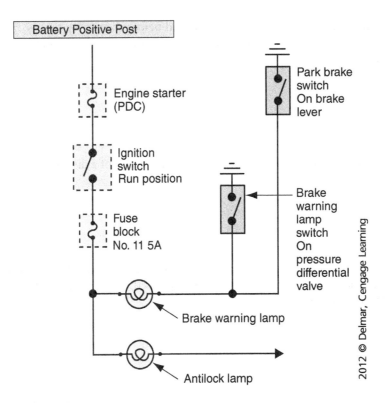

40. The brake warning lamp pictured above will occasionally come on then later go off. Which of the following could be the cause?

    A. A misadjusted park brake switch
    B. Water contaminated brake fluid
    C. Worn parking brakes
    D. Worn brake drums

41. The brake pedal has a pulsation when the brakes are applied. Technician A says the drum may be out of round. Technician B says the drum may be glazed. Who is correct?

    A. A
    B. B
    C. Both A and B
    D. Neither A nor B

42. The technician makes one measurement on a disc brake rotor with a micrometer. Which of the following is being measured?

    A. Brake rotor diameter
    B. Brake rotor thickness
    C. Brake rotor lateral runout
    D. Brake rotor thickness variation

43. The ABS controller has stored a code for the left front wheel speed sensor. All of the following could be the cause EXCEPT:

    A. A wheel speed sensor resistance reading of 0.2 ohms.
    B. Mismatched tire diameters.
    C. A loose wheel bearing on the left front.
    D. A loose wheel bearing on the right front.

Section 5 Sample Preparation Exams   Brakes (A5)

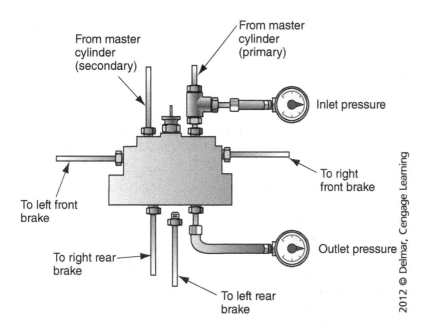

44. During a pressure test both gauges read lower than normal. Technician A says low fluid level in the primary reservoir could be the cause. Technician B says low fluid level in the secondary reservoir could be the cause. Who is correct?

    A. A
    B. B
    C. Both A and B
    D. Neither A nor B

45. While testing the vacuum brake booster the technician finds that if the engine is started with pressure applied to the brake pedal the pedal does not move. Which of the following is indicated by this test result?

    A. A restricted vacuum hose
    B. A leaking master cylinder
    C. A bypassing master cylinder
    D. A normally operating booster

PREPARATION EXAM 6

1. What would a C-clamp be used for during disc brake service?

    A. Seating the pads on the rotor
    B. Seating the caliper on the guides
    C. Retracting the piston in the bore
    D. Compressing the pads into the piston

2. The red brake warning light illuminates when the brake pedal is depressed. Which of the following is the LEAST LIKELY cause?

    A. Air in the front brake system
    B. Air in the rear brake system
    C. Low brake fluid level
    D. Faulty ABS pump assembly

3. Which of the following would be considered the most normal brake fluid service interval?
   A. 1 year/15,000 miles
   B. 3 year/36,000 miles
   C. 6 years/72,000 miles
   D. 10 years/100,000 miles

4. The dimension stamp on a rotor is 1.18" (30.00mm). The left front rotor measures 1.17" (31.75mm). The right front rotor measures 1.29" (29.72). Which of the following is true?
   A. The left rotor needs to be replaced.
   B. The right rotor needs to be replaced.
   C. Both rotors need to be replaced.
   D. Neither rotor needs to be replaced.

5. The brake fluid in a vehicle is black. Which of the following is the most likely cause?
   A. The brake fluid is contaminated with power steering fluid.
   B. The brake fluid is contaminated with water.
   C. Black is a normal brake fluid color.
   D. The brake fluid needs to be replaced with new.

6. Technican A says a 1/4" drill bit may be needed to check the parking brake adjustment on some vehicles. Technician B says a hose clamp can be used to release the parking brake cable from the backing plate. Who is correct?
   A. A
   B. B
   C. Both A and B
   D. Neither A nor B

7. The brake pedal vibrates when applied on a vehicle with a hydro-boost brake booster. Which of the following is the most likely cause?
   A. A restricted booster supply hose
   B. A loose power steering pump belt
   C. A restricted front brake line
   D. A restricted rear brake line

8. When bleeding the brakes using the brake pedal, which of the following is true?
   A. A hose should be attached to the bleeder screw and submerged in brake fluid.
   B. The bleeder screw should be closed throughout the downward stroke of the brake pedal.
   C. The bleeder screw should be loosened two complete turns.
   D. The brake pedal should be stroked as quickly as possible.

9. Technician A says the traction control system can signal the engine control module to reduce engine torque. Technician B says the traction control system can apply the vehicle brakes. Who is correct?
   A. A
   B. B
   C. Both A and B
   D. Neither A nor B

10. The specification stamped on the drum is 8.125" (206.38mm). Which measurement below would indicate a re-useable drum?

    A. 8.025" (203.84mm)
    B. 8.126" (206.38mm)
    C. 8.225" (208.92mm)
    D. 8.325" (211.45mm)

11. The rear brakes tend to lock during hard braking. Which of the following could be the cause?

    A. Metering valve
    B. Proportioning valve
    C. Quick take up valve
    D. Pressure differential valve

12. Which of the following best describes a four-wheel antilock (4WAL) antilock brake system?

    A. It is a four-wheel antilock system used only on four-wheel drive vehicles.
    B. It is a four-wheel antilock system used only on all-wheel drive vehicles.
    C. It is a four-wheel antilock system used only on pick-up trucks.
    D. It is a four-wheel antilock system used on a variety of vehicles.

13. Rotor runout is checked with a dial indicator and found to be out of specification. The rotor is removed, rotated 180 degrees, and reinstalled. The runout is checked again and is still out of specification. The excessive runout is in the same location on the hub. Which of the following is indicated?

    A. The measurement is being performed incorrectly.
    B. The dial indicator is faulty.
    C. The runout is in the hub.
    D. The runout is in the rotor.

14. Which of the following would cause a drum to have hard spots?

    A. A broken return spring
    B. Air in the hydraulic system
    C. Missing brake shoe lining
    D. A frozen wheel cylinder

15. When the brakes are applied at low speed on a slippery surface the front brakes lock. Which of the following could be the cause?

    A. Metering valve
    B. Proportioning valve
    C. Quick take up valve
    D. Pressure differential valve

16. The parking brake warning light switch is misadjusted. Which light on the dash will most likely illuminate?

    A. Red brake warning light
    B. Amber brake warning light
    C. ABS light
    D. Traction control (TRAC) light

17. Which of the following is the correct disposal method for used brake fluid?
    A. Pour it down the drain.
    B. Pour it in with used oil to be recycled.
    C. Pour it on the ground.
    D. Save it in a separate container for recycling.

18. Technician A says machining the rotor on the vehicle helps to compensate for hub runout. Technician B says machining the rotor on the vehicle helps to compensate for wheel runout. Who is correct?
    A. A
    B. B
    C. Both A and B
    D. Neither A nor B

19. Which of the following best describes a three channel ABS system?
    A. The rear wheels share a circuit.
    B. The front wheels share a circuit.
    C. The left side shares a circuit.
    D. The right side shares a circuit.

20. Technician A says DOT 5 brake fluid is clear. Technician B says DOT 3 brake fluid is purple. Who is correct?
    A. A
    B. B
    C. Both A and B
    D. Neither A nor B

21. Technician A says the red brake warning light is turned on by the parking brake warning light switch. Technician B says the red brake warning light is turned on by the pressure differential switch. Who is correct?
    A. A
    B. B
    C. Both A and B
    D. Neither A nor B

22. What is meant by "indexing" a rotor?
    A. Matching the high spot on the rotor and low spot on the hub
    B. Matching the low spot on the rotor and the low spot on the hub
    C. Matching the high spot on the rotor and the high spot on the hub
    D. Matching the widest spot on the rotor to the widest spot on the hub

23. On a vehicle with disc brakes, the outside pad is worn much more than the inside pad. Which of the following could be the cause?
    A. Sticking caliper slides
    B. Collapsed brake line
    C. A sticking metering valve
    D. Swollen brake line

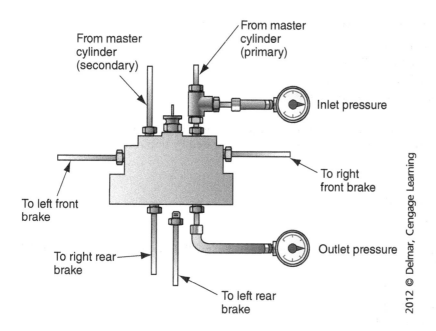

24. After a brake application, the inlet pressure gauge drops to normal; however, the outlet pressure gauge remains higher than normal. Technician A says a faulty combination valve could be the cause. Technician B says a faulty antilock braking system (ABS) modulator valve could be the cause. Who is correct?

   A. A
   B. B
   C. Both A and B
   D. Neither A nor B

25. A parking brake drags after application and release. Technician A says the cable may be rusty. Technician B says the brake drum may be bell mouthed. Who is correct?

   A. A
   B. B
   C. Both A and B
   D. Neither A nor B

26. Technician A says some vacuum brake booster systems used on diesel engines will have an engine driven vacuum pump. Technician B says some vacuum brake booster systems used on gasoline engines will have an engine driven vacuum pump. Who is correct?

   A. A
   B. B
   C. Both A and B
   D. Neither A nor B

27. Which of the following tools would most likely be used to measure wheel bearing end-play?

   A. Outside micrometer
   B. Inside micrometer
   C. Dial indicator
   D. Feeler gauge

28. Which of the following would cause a grooved drum?

    A. A dragging shoe
    B. Air in the hydraulic system
    C. Missing brake shoe lining
    D. A frozen parking brake cable

29. Technicain A says excessive rotor runout can be caused by a warped disc brake rotor. Technician B says excessive rotor runout can be caused by hub runout. Who is correct?

    A. A
    B. B
    C. Both A and B
    D. Neither A nor B

30. When constant pressure is held on the brake pedal it will slowly fade to the floor. Which is the most likely cause?

    A. Air in the brake fluid
    B. Stuck caliper piston
    C. Stuck wheel cylinder
    D. Internally leaking master cylinder

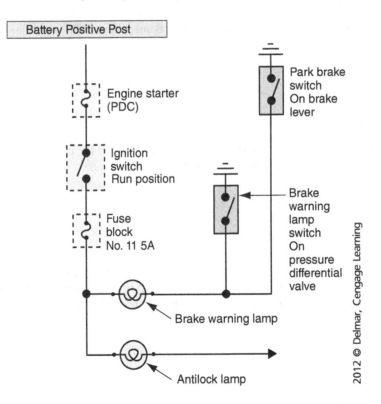

31. The brake warning lamp pictured above stays on any time the ignition switch is on. Technician A says this could be caused by a stuck closed parking brake switch. Technician B says this could be caused by a stuck closed brake warning lamp switch. Who is correct?

    A. A
    B. B
    C. Both A and B
    D. Neither A nor B

32. The brake pedal is slow to return when released on a vehicle with a hydro-boost brake booster. Which of the following is the most likely cause?
    A. A restricted booster supply hose
    B. A restricted booster return hose
    C. A restricted front brake line
    D. A restricted rear brake line

33. Technician A says some disc brake rotors are directional and can be installed only on one side of a car. Technician B says rotor thickness variation and parallelism refer to the same measurement. Who is correct?
    A. A
    B. B
    C. Both A and B
    D. Neither A nor B

34. The wheels speed sensor is being checked with an ohmmeter. When the technician measures between one of the wires and chassis ground the ohmmeter shows 0.2 ohms. This indicates:
    A. The sensor is open.
    B. The sensor is shorted to voltage.
    C. The sensor is shorted to chassis ground.
    D. The sensor is functioning correctly.

35. Technician A says the coil spring wrapped around a section of steel brake line is to help dissipate heat. Technician B says the spring wrapped around a section of steel brake line is to protect the line from abrasion wear. Who is correct?
    A. A
    B. B
    C. Both A and B
    D. Neither A nor B

36. Technician A says the RWAL system may be deactivated on some vehicles when they are put in four-wheel drive mode. Technician B says some G force sensors are checked by measuring resistance while tilting the sensor. Who is correct?
    A. A
    B. B
    C. Both A and B
    D. Neither A nor B

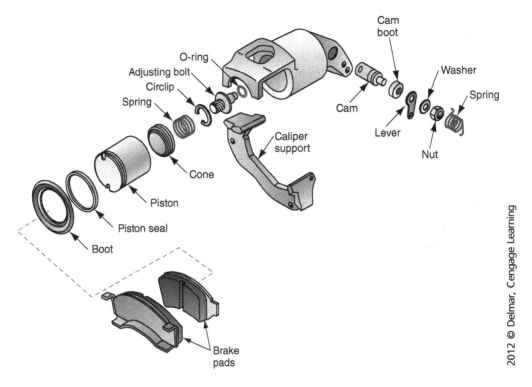

37. Refer to the illustration above. The parking brake works correctly; however, the brake does not operate when the service brake pedal is pushed. Technician A says the problem could be a stuck adjuster. Technician B says the problem could be a seized cam. Who is correct?

   A. A
   B. B
   C. Both A and B
   D. Neither A nor B

38. Rotor runout is checked with a dial indicator and found to be out of specification. The rotor is removed, rotated 180 degrees, and reinstalled. The runout is checked again and is still out of specification. The excessive runout is in the same location on the rotor. Which of the following is most likely indicated?

   A. The measurement is being performed incorrectly.
   B. The dial indicator is faulty.
   C. The runout is in the hub.
   D. The runout is in the rotor.

39. Which of the following would be the most likely cause of an ABS code?

   A. A grounded pad wear sensor
   B. Air in the brake fluid
   C. A cracked tone ring
   D. A malfunctioning proportioning valve

40. Brake fluid is leaking from the master cylinder reservoir cap. The seal is swollen and deformed. Which of the following could be the cause?

    A. Brake fluid contaminated with water
    B. Brake fluid contaminated with oil
    C. Brake fluid contaminated with metals
    D. Brake fluid has been overheated

41. Rotor runout is checked with a dial indicator and found to be out of specification. The rotor is removed, rotated 180 degrees, and reinstalled. The runout is checked again and found to be within specification. Which of the following is indicated?

    A. The rotor should be replaced.
    B. The hub should be replaced.
    C. The brake can be assembled in this position.
    D. The rotor should be turned 90 degrees and installed.

42. Which of the following is the hardest to remove from a drum with a drum brake lathe?

    A. Bell mouth
    B. Out of round
    C. Hard spots
    D. Grooving

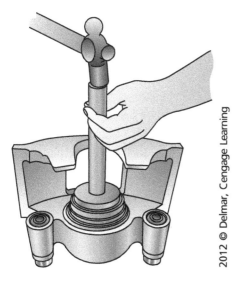

43. Technician A says the pin bushings are being installed in the illustration above. Technician B says the boot is being removed in the above illustration. Who is correct?

    A. A
    B. B
    C. Both A and B
    D. Neither A nor B

44. When replacing the master cylinder, which of the following is true?
    A. The booster should also be replaced.
    B. The disc brake calipers should also be replaced.
    C. The drum brake wheel cylinders must also be replaced.
    D. The master cylinder should be bench bled prior to installation.

45. A technician is testing the accumulator on a hydro-boost brake booster. Which of the following is a correct test procedure?
    A. Apply the brakes and hold, then start the engine.
    B. Apply the brakes and hold, then shut the engine off.
    C. Turn the engine off and apply the brakes.
    D. Apply the brakes, turn the engine off, release the brakes and restart the engine.

SECTION 6

Answer Keys and Explanations

Included in this section are the answer keys for each preparation exam, followed by individual, detailed answer explanations and a reference identifying the designated task area being assessed by each specific question. This additional reference information may prove useful if you need to refer back to the task list located in Section 4 of this book for additional support.

PREPARATION EXAM 1—ANSWER KEY

1. A	21. A	41. B
2. C	22. B	42. B
3. D	23. B	43. A
4. C	24. A	44. D
5. C	25. A	45. A
6. D	26. A	
7. A	27. D	
8. D	28. A	
9. C	29. C	
10. A	30. C	
11. B	31. A	
12. A	32. A	
13. D	33. C	
14. C	34. A	
15. B	35. A	
16. B	36. B	
17. C	37. C	
18. B	38. A	
19. D	39. B	
20. A	40. C	

Delmar, Cengage Learning ASE Test Preparation

PREPARATION EXAM 1—EXPLANATIONS

TASK B.4

1. A drum is chattering on the brake lathe. This could be the result of:
   A. Failure to install the dampening belt.
   B. Cutting too slowly.
   C. Dull cutting blades.
   D. Cutting speed too fast.

   Answer A is correct. A missing dampening belt could cause the drum to vibrate while on the lathe. This vibration can cause the drum to bounce on the cutting tip and make a chattering noise.

   Answer B is incorrect. Cutting too slowly while not cause chatter.

   Answer C is incorrect. Dull cutting bits leave a rough finish in the drum but usually do not cause chatter.

   Answer D is incorrect. A fast cutting speed will cause the drum to be threaded.

TASK D.4

2. A vehicle equipped with a hydro-boost brake system has a hard pedal with little braking action. Which of the following could be the cause?
   A. Low brake fluid
   B. Air in the brake fluid
   C. Low power steering fluid
   D. An overtightened power steering belt

   Answer A is incorrect. Low brake fluid would cause a low pedal.

   Answer B is incorrect. Air in the brake fluid would cause a low, spongy pedal.

   Answer C is correct. Low power steering fluid can cause the booster to provide very little if any boost.

   Answer D is incorrect. An overtightened belt may prematurely wear the belt; however, it would not cause the booster to be weak.

TASK C.2

3. Which of the following could cause excessive noise while braking?
   A. Worn wheel cylinders
   B. Worn caliper pistons
   C. Seized caliper slides
   D. Missing clips

   Answer A is incorrect. Worn wheel cylinders can cause a leak, not a noise.

   Answer B is incorrect. Worn caliper pistons normally cause a leak.

   Answer C is incorrect. Seized caliper slides would not cause a noise.

   Answer D is correct. Missing anti-rattle clips could cause excessive noise while braking.

4. A vehicle equipped with a "hill holding" option has an ABS code. Which tool would most likely be used to check the wheel speed sensor?

   A. AC volt meter
   B. Ohm meter
   C. Scan tool
   D. Ammeter

   Answer A is incorrect. A vehicle with a hill holding option will not have analog wheel speed sensors.

   Answer B is incorrect. An ohmmeter cannot be used to check a digital wheel speed sensor.

   Answer C is correct. A scan tool or oscilloscope can be used to test a signal from a digital wheel speed sensor.

   Answer D is incorrect. An ammeter would not be used to test a wheel speed sensor.

5. The brake system has a soft, sinking pedal with the engine running. The brake fluid level is at the correct height. Which of the following could be the cause?

   A. A restricted vacuum hose
   B. A leaking master cylinder
   C. A bypassing master cylinder
   D. A stuck open vacuum check valve

   Answer A is incorrect. A restricted vacuum hose would cause a hard pedal with poor braking performance.

   Answer B is incorrect. A leaking master cylinder would cause a soft, sinking pedal; however, the brake fluid would be low due to the leak.

   Answer C is correct. A bypassing master cylinder would cause a low pedal and would not cause a loss of fluid.

   Answer D is incorrect. A stuck open vacuum check valve would cause no vacuum assist after the engine was shut off.

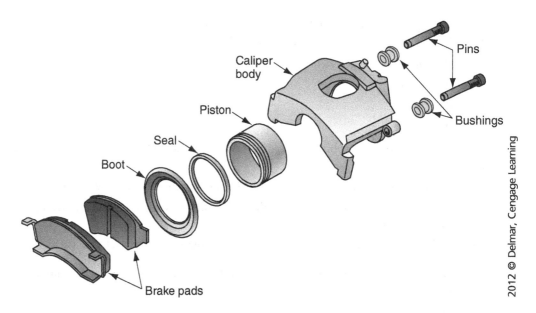

TASK C.11

6. Refer to the figure above. Technician A says the pins should be driven in with a brass hammer. Technician B says the pins should be lubricated with brake fluid. Who is correct?

   A. A
   B. B
   C. Both A and B
   D. Neither A nor B

   Answer A is incorrect. A brass hammer should not be used on this style of pin.

   Answer B is incorrect. Brake fluid is not used as slide lubricant.

   Answer C is incorrect. Neither technician is correct.

   Answer D is correct. Neither technician is correct. This style of pin should be installed with a torx or hex head socket.

TASK D.4

7. Technician A says a hydro-boost system can use the power steering pump to operate the booster. Technician B says a hydro-boost system can use engine vacuum to operate the booster. Who is correct?

   A. A
   B. B
   C. Both A and B
   D. Neither A nor B

   Answer A is correct. Only Technician A is correct. A hydro-boost system uses hydraulic pressure to create boost.

   Answer B is incorrect. Vacuum is not used in a hydro-boost system. Vacuum is used in a vacuum booster.

   Answer C is incorrect. Only Technician A is correct.

   Answer D is incorrect. Technician A is correct.

8. A disc brake caliper is dragging. All of the following could be the cause EXCEPT:

   A. Binding pins.
   B. Sticking piston.
   C. Binding caliper.
   D. Worn brake pads.

   TASK C.2

   Answer A is incorrect. Binding pins will cause the caliper to hang and the brakes to drag.

   Answer B is incorrect. A sticking piston will not retract and cause the brakes to drag.

   Answer C is incorrect. A binding caliper will not release and cause the brakes to drag.

   Answer D is correct. Worn brake pads will not cause the brakes to drag.

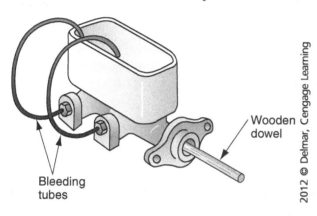

9. Which operation is being performed in the illustration above?

   A. Wheel cylinder bleeding
   B. Brake caliper bleeding
   C. Master cylinder bleeding
   D. Master cylinder pressure test

   TASK A.1.6

   Answer A is incorrect. This is not a wheel cylinder.

   Answer B is incorrect. This is not a caliper.

   Answer C is correct. The master cylinder is being bleed prior to installation.

   Answer D is incorrect. A pressure gauge would be required to check master cylinder pressure.

10. The rear wheels of a front wheel drive car lockup during heavy braking. Technician A says this could be caused by a faulty proportioning valve. Technician B says this could be caused by a faulty brake booster. Who is correct?

    A. A only
    B. B only
    C. Both A and B
    D. Neither A nor B

    TASK A.3.2

    Answer A is correct. Technician A is correct. A faulty proportioning valve could cause the pressure to the rear wheels to be too high and lock the wheels during hard braking.

    Answer B is incorrect. A faulty brake booster would affect all wheels and not just the rear.

    Answer C is incorrect. Only Technician A is correct.

    Answer D is incorrect. Technician A is correct.

Section 6 Answer Keys and Explanations — Brakes (A5)

TASK E.1

11. There is a loud growling noise only when the vehicle is steered to the right while driving. Which of the following could be the cause?

    A. Worn left front brakes
    B. Worn wheel bearing
    C. Worn right front brakes
    D. Sticking brake caliper

    Answer A is incorrect. Worn left front brakes would not cause a noise only when steered to the right.

    Answer B is correct. This is a typical wheel bearing noise.

    Answer C is incorrect. Worn right front brakes would not cause a noise only when steered to the right.

    Answer D is incorrect. A sticking caliper would not cause a noise only when steered to the right.

TASK B.5

12. A spongy pedal can be the result of:

    A. Leaking wheel cylinders.
    B. Broken return springs.
    C. Contaminated linings.
    D. Seized wheel cylinder pistons.

    Answer A is correct. A spongy pedal can result from air in the hydraulic system. Leaking wheel cylinders could allow air to get into the hydraulic system and result in a spongy pedal.

    Answer B is incorrect. Broken return springs could cause the drum brakes to drag.

    Answer C is incorrect. Contaminated linings could cause the brake shoes to stick and slip, which would cause a chattering condition.

    Answer D is incorrect. Seized wheel cylinders could cause a hard pedal condition with a lack of braking power.

TASK F.8

13. A vehicle has a wheel speed sensor diagnostic trouble code. Which of the following is the LEAST LIKELY cause?

    A. Faulty wheel speed sensor
    B. Cracked wheel speed sensor reluctor ring
    C. A tire with the incorrect height
    D. A tire with the incorrect width

    Answer A is incorrect. A faulty wheel speed sensor is a likely cause of a wheel speed sensor diagnostic trouble code.

    Answer B is incorrect. A cracked wheel speed sensor reluctor ring can cause a faulty wheel speed sensor signal to be generated and set a diagnostic trouble code.

    Answer C is incorrect. Mismatched tire height can cause the speed signals to vary between the different sensors and set a diagnostic trouble code.

    Answer D is correct. The width of the tire will not affect the wheel speed sensor signal.

14. A vehicle has a squeak only when applying the brakes. Which of the following could be the cause?

    A. Aerated brake fluid
    B. Water contaminated brake fluid
    C. Worn brake pads
    D. Worn wheel bearings

    Answer A is incorrect. Aerated fluid causes a low pedal, not squeaks.

    Answer B is incorrect. Water in the brake fluid will cause the fluid to boil at low temperatures and a soft pedal.

    Answer C is correct. Worn pads can cause a squeak when applying the brakes.

    Answer D is incorrect. Worn wheel bearings can cause a squeak, however it would not be only when braking.

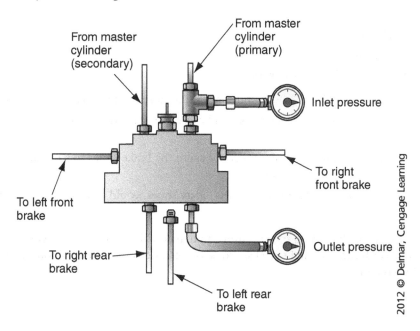

15. During a moderate brake pedal application, with the engine running, which of the following would be considered a normal pressure reading on the inlet pressure gauge shown in the figure above?

    A. 100 psi
    B. 900 psi
    C. 3000 psi
    D. 5000 psi

    Answer A is incorrect. This would be too low.

    Answer B is correct. This would be normal.

    Answer C is incorrect. This would be high for moderate brake pedal pressure.

    Answer D is incorrect. This would be extremely high for moderate pedal pressure.

16. Which of the following would be the most normal end-play specification for a set of tapered roller front wheel bearings?

   A. 0.0001″–0.0005″ (0.00254mm–0.0127mm)
   B. 0.0010″–0.0050″ (0.0254mm–0.127mm)
   C. 0.0100″–0.0500″ (0.254mm–1.27mm)
   D. 0.1000″–0.5000″ (2.54mm–12.7mm)

   Answer A is incorrect. This would be too little clearance.
   Answer B is correct. This is the correct specification.
   Answer C is incorrect. This would be too much clearance.
   Answer D is incorrect. This would be too much clearance.

17. Brake chatter can be caused by:

   A. Seized star-wheel adjuster.
   B. Seized wheel cylinder.
   C. Contaminated linings.
   D. Switched primary and secondary shoes.

   Answer A is incorrect. Seized star-wheel adjusters would cause the brakes not to self adjust.
   Answer B is incorrect. A seized wheel cylinder will cause the brake on that wheel end to be inoperative.
   Answer C is correct. Contaminated linings could cause the brake shoes to stick and slip, which would cause a chattering condition.
   Answer D is incorrect. Switched primary and secondary shoes would cause reduced braking power.

18. Technician A says an analog wheel speed sensor can detect which direction the wheel is turning. Technician B says a digital wheel speed sensor can detect which direction the wheel is turning. Who is correct?

   A. A
   B. B
   C. Both A and B
   D. Neither A nor B

   Answer A is incorrect. An analog sensor generates an ac voltage signal and does not indicate which direction the wheel is turning.
   Answer B is correct. Only Technician B is correct. A digital wheel speed sensor can indicate direction by detecting narrow and wide slots on the reluctor ring.
   Answer C is incorrect. Only Technician B is correct.
   Answer D is incorrect. Technician B is correct.

19. All of the following are the result of brake pad burnishing EXCEPT:

    A. Forms the new pads to the rotor.
    B. Cures the materials in the friction material.
    C. Provides for better braking.
    D. Cures the metal in the rotor.

    Answer A is incorrect. Burnishing forms the pads to the shape of the rotor.

    Answer B is incorrect. Burnishing cures the materials in the friction material.

    Answer C is incorrect. Burnishing provides for better braking.

    Answer D is correct. Brake rotors do not need the metal cured.

20. When bench bleeding a master cylinder which statement is true?

    A. New brake fluid should be used.
    B. Brake cleaning fluid should be used.
    C. Mineral spirits should be used.
    D. Mineral oil should be used.

    Answer A is correct. New brake fluid should be used when bleeding the master cylinder.

    Answer B is incorrect. Brake cleaning fluid should not be used. Brake cleaning fluid is used to clean brake components.

    Answer C is incorrect. Mineral spirits should never be used on brake parts which will contact brake fluid.

    Answer D is incorrect. Mineral oil should never be used on brake parts which will contact brake fluid.

21. The technician is preparing to do a brake job on a vehicle equipped with integral ABS. Which of the following should be done?

    A. Pump the pedal 25–40 times
    B. Remove the fuses for the ABS controller
    C. Apply the parking brake
    D. Remove the fuse for the ignition system

    Answer A is correct. The pedal should be pumped 25–40 times to bleed all the pressure from the ABS accumulator.

    Answer B is incorrect. Fuse removal is not necessary.

    Answer C is incorrect. If the parking brake is applied the shoes cannot be removed.

    Answer D is incorrect. The fuse does not need to be removed.

Section 6 Answer Keys and Explanations — Brakes (A5)

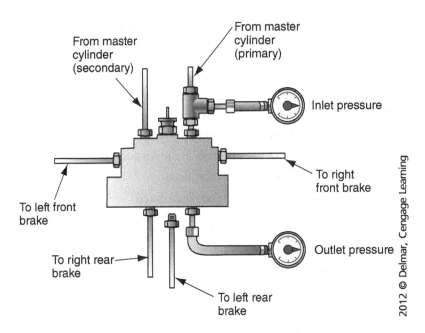

TASK A.3.2

22. Refer to the illustration above. Technician A says pressure is being tested at the antilock braking system (ABS) modulator valve. Technician B says pressure is being tested at the combination valve. Who is correct?

   A. A only
   B. B only
   C. Both A and B
   D. Neither A nor B

   Answer A is incorrect. This is not a modulator valve.

   Answer B is correct. Only Technician B is correct. This is a combination valve.

   Answer C is incorrect. Only Technician B is correct.

   Answer D is incorrect. Technician B is correct.

TASK E.1

23. A customer has a noise concern coming from the rear of the vehicle. All of the following could be the cause EXCEPT:

   A. Worn rear wheel bearings.
   B. Rear tires out of balance.
   C. Dragging brake shoes.
   D. Binding parking brake cable.

   Answer A is incorrect. Worn rear wheel bearings can cause noise from the rear.

   Answer B is correct. Rear tires out of balance would cause a vibration but not a noise.

   Answer C is incorrect. Dragging brake shoes can cause noise from the rear.

   Answer D is incorrect. A binding parking brake cable can cause the brakes to drag, thus noise.

Section 6 Answer Keys and Explanations

Brakes (A5)

24. A vehicle comes back into the shop with a brake pulsation after having the rotors replaced. Which of the following is the most likely cause?

    A. Over torqued lug nuts
    B. Under torqued lug nuts
    C. Sticking slides
    D. Incorrect brake fluid installed

    TASK C.14

    Answer A is correct. A wheel torque that is too tight could warp the rotors. This could result in pedal pulsation.

    Answer B is incorrect. Under torqued nuts would cause a loose wheel not a warped rotor.

    Answer C is incorrect. Sticking slides would cause a pulling condition.

    Answer D is incorrect. Using the incorrect brake fluid will not cause the rotor to warp.

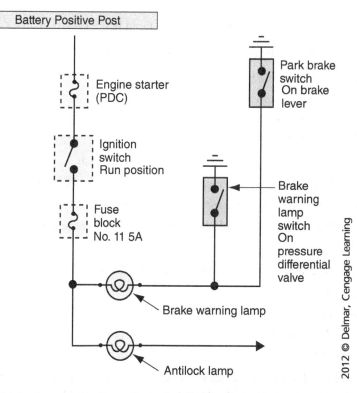

25. The brake warning lamp illustrated in the figure above does not illuminate when the parking brake is applied. It operates normally otherwise. Which of the following could be the cause?

    A. An open ground at the parking brake switch
    B. An open ground at the master cylinder
    C. An open fuse 11
    D. An open engine starter fuse

    TASK E.5

    Answer A is correct. An open ground at the parking brake switch would prevent the light from illuminating when the parking brake is activated.

    Answer B is incorrect. The master cylinder switch is not completing the circuit when the parking brake is applied.

    Answer C is incorrect. The circuit works normally otherwise; therefore, the fuse must be OK. An open fuse would prevent circuit operation under any circumstances.

    Answer D is incorrect. The circuit works normally otherwise; therefore, the fuse must be OK. An open fuse stops all current flow, making the circuit totally inoperative.

TASK F.2

26. A customer is concerned that on dry pavement, their vehicle will occasionally have a rapid brake pulsation similar to the feel of an ABS stop. Which of the following is the most likely cause?

   A. Faulty wheel speed sensor
   B. Faulty master cylinder
   C. Faulty ABS control module
   D. Faulty brake booster

   Answer A is correct. A faulty wheel speed sensor can give false information which may cause ABS activation when it is not needed.

   Answer B is incorrect. The master cylinder will not cause a rapid brake pulsation.

   Answer C is incorrect. While the ABS modulator valve may possibly cause this condition it is not the most likely.

   Answer D is incorrect. A brake booster does not cause a rapid brake pulsation.

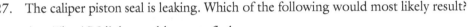

TASK A.4.1

27. The caliper piston seal is leaking. Which of the following would most likely result?

   A. The ABS light would start to flash.
   B. The traction control light would start to flash.
   C. The brake pedal would become very hard.
   D. The brake pedal would become very soft.

   Answer A is incorrect. A fluid leak will not cause the ABS light to flash.

   Answer B is incorrect. A fluid leak would not cause the traction control light to flash.

   Answer C is incorrect. A fluid loss does not cause a hard pedal.

   Answer D is correct. The brake pedal would become soft because as the fluid got low, air would enter the system.

TASK C.1

28. A vehicle pulls right while braking. The most likely cause would be:

   A. Seized left caliper slides.
   B. Faulty metering valve.
   C. Seized right rear wheel cylinder.
   D. Faulty master cylinder.

   Answer A is correct. Seized left caliper slides would cause a lack of braking force on the left side. The higher braking force on the right side would cause the vehicle to pull to the right.

   Answer B is incorrect. A faulty metering valve will not cause a pull condition.

   Answer C is incorrect. A seized right rear wheel cylinder would not cause a vehicle to pull to the right.

   Answer D is incorrect. A faulty master cylinder can cause a brake performance problem, but it does not normally cause a pull.

29. The pressure differential valve lights the red brake warning light when:

    A. The parking brake pedal is depressed.
    B. There is an ABS code.
    C. There is a pressure imbalance.
    D. The brake fluid is low.

    TASK A.3.2

    Answer A is incorrect. The pressure differential switch lights the light when there is a pressure differential.

    Answer B is incorrect. The amber light is the ABS light.

    Answer C is correct. A pressure differential will cause the switch to illuminate the light.

    Answer D is incorrect. The level switch indicates brake fluid level.

30. A vehicle equipped with a hydro-boost brake system is being diagnosed for a high, hard pedal. Which of the following tools could be used?

    A. A vacuum gauge
    B. A black light
    C. A power steering pressure tester
    D. A vacuum pump

    TASK D.4

    Answer A is incorrect. This is not a vacuum booster.

    Answer B is incorrect. A black light would not help find the cause of low booster action.

    Answer C is correct. A power steering pressure tester would be used to test the pressure and flow from the pump.

    Answer D is incorrect. This is not a vacuum booster.

31. A vehicle has a low, firm brake pedal. Which of the following would most likely be the cause?

    A. Improperly adjusted brake shoes
    B. Leaking wheel cylinders
    C. Excessive rotor runout
    D. Out of round drums

    TASK B.1

    Answer A is correct. Improperly adjusted shoes can cause the wheel cylinder pistons to travel further, thus causing a low, firm pedal.

    Answer B is incorrect. While leaking wheel cylinders could cause a low brake pedal, the pedal would also be spongy, not firm.

    Answer C is incorrect. Rotor runout will cause a pedal pulsation concern.

    Answer D is incorrect. Out of round drums would cause a pulsation complaint instead of a low pedal complaint.

Section 6 Answer Keys and Explanations   Brakes (A5)

TASK C.13

32. The brake fluid reservoir is low on fluid. Which of the following is the most likely cause?
    A. Worn pads
    B. Worn shoes
    C. Bypassing master cylinder
    D. Leaking vacuum check valve

    Answer A is correct. Worn front brake pads will cause the caliper piston to be pushed out further to take up for the wear. More fluid will be stored in the caliper as a result.

    Answer B is incorrect. Worn shoes are adjusted with an adjuster, not by adding fluid to the wheel cylinders.

    Answer C is incorrect. A bypassing master cylinder will cause a low falling pedal, not low fluid.

    Answer D is incorrect. A leaking vacuum check valve will cause a no boost reserve condition.

TASK E.1

33. A customer says their car makes more noise on asphalt than on concrete road surfaces. Which of the following could be the cause?
    A. Front wheel bearing wear
    B. Rear wheel bearing wear
    C. Aggressive tire tread patterns
    D. Worn brake pads

    Answer A is incorrect. Wheel bearing noises will change with load not road surface finish.

    Answer B is incorrect. Wheel bearing noises will change with load not road surface finish.

    Answer C is correct. Tires with very aggressive tread patterns are noisy. This noise will vary with different types of finish on the road surface.

    Answer D is incorrect. Worn brake pads will not cause tire noise to change while driving.

TASK A.4.1

34. Which of the following would most likely cause a brake fluid leak on a front disc brake caliper assembly?
    A. Piston seal
    B. Piston boot
    C. Caliper slide pins
    D. Caliper slide bushings

    Answer A is correct. A faulty piston seal could cause a leak.

    Answer B is incorrect. A boot would not cause a leak. It may cause a stuck piston.

    Answer C is incorrect. The pins would not cause a leak. They may cause a stuck caliper.

    Answer D is incorrect. Bushings would not cause a leak. They may cause uneven pad wear.

35. All of the following could cause a brake pedal pulsation EXCEPT:
    A. Thin brake pad linings.
    B. Excessive lug nut torque.
    C. Excessive rotor thickness variation.
    D. Rust build-up between the hub and rotor.

    TASK C.2

    Answer A is correct. Worn brake pad linings can cause noise, but will not cause a brake pedal pulsation.

    Answer B is incorrect. Excessive lug nut torque can warp a rotor and cause a brake pedal pulsation.

    Answer C is incorrect. Excessive rotor thickness variation could cause a brake pedal pulsation.

    Answer D is incorrect. Rust build-up between the rotor and hub would cause excessive lateral run-out and cause a brake pedal pulsation.

36. There is a loud growling noise while driving. The noise gets louder when vehicle speed increases. Which of the following could be the cause?
    A. Worn rear brake shoes
    B. Worn wheel bearing
    C. Worn ball joint
    D. Worn front brake pads

    TASK E.1

    Answer A is incorrect. Brake shoes will not cause a growling noise.

    Answer B is correct. This is a typical wheel bearing diagnosis.

    Answer C is incorrect. A worn ball joint can cause a knocking noise in certain suspension travel positions. However, it would not cause a growling noise.

    Answer D is incorrect. Worn front brake pads do not make a growling noise.

37. Which of the following would be the correct way to handle brake fluid?
    A. Because brake fluid does not have an expiration date, purchase in the largest containers possible for economy.
    B. Store in a container vented to the atmosphere.
    C. Store in small, tightly sealed containers.
    D. Store in a temperature controlled environment, preferably refrigerated.

    TASK A.4.3

    Answer A is incorrect. Brake fluid does not have an expiration date, but is hygroscopic, so it should be purchased in small containers which will be used quickly.

    Answer B is incorrect. The container should be sealed, not vented.

    Answer C is correct. This is the correct procedure.

    Answer D is incorrect. Brake fluid does not need to be stored in a temperature-controlled environment.

TASK F.7

38. Technician A says rust can cause a faulty wheel speed sensor signal. Technician B says a defective brake switch can cause a faulty wheel speed signal. Who is correct?

A. A
B. B
C. Both A and B
D. Neither A nor B

Answer A is correct. Only Technician A is correct. Rust can cause the wheel speed sensor to be pushed out of the housing it sits in, and result in a faulty wheel speed sensor signal.

Answer B is incorrect. A defective brake switch will not cause a faulty wheel speed signal.

Answer C is incorrect. Only Technician A is correct.

Answer D is incorrect. Technician A is correct.

TASK D.4

39. A vehicle equipped with a hydro-boost brake system has a low, soft pedal with little braking action. Which of the following could be the cause?

A. Weak power steering pump
B. Air in the brake fluid
C. Air in the power steering fluid
D. Low intake manifold vacuum

Answer A is incorrect. A weak pump would cause a high, hard pedal.

Answer B is correct. Air in the brake fluid would cause a low, soft pedal.

Answer C is incorrect. Air in the power steering fluid can cause the booster to provide very little if any boost. This would cause a hard pedal.

Answer D is incorrect. Low manifold vacuum would not affect this system.

TASK E.2

40. A set of tapered roller bearings are being serviced. Which of the following is true concerning the correct tightening procedure?

A. Tighten the nut to 75 ft lbs.
B. Tighten the nut to spec and back off one turn.
C. The final check should be made with a dial indicator.
D. The wheels/tires must be on the ground.

Answer A is incorrect. Seventy-five ft lbs would apply too much preload to the bearing.

Answer B is incorrect. This would allow too much freeplay.

Answer C is correct. The final check should be 0.001"–0.005" (0.025mm–0.127mm).

Answer D is incorrect. The wheel end must be rotated during the tightening procedure.

Section 6 Answer Keys and Explanations

Brakes (A5)

41. The brake system has a low, soft pedal with the engine running. The brake fluid is low in the primary reservoir. Which of the following could be the cause?

    A. A restricted vacuum hose
    B. A leaking master cylinder
    C. A bypassing master cylinder
    D. A stuck open vacuum check valve

    TASK C.13

    Answer A is incorrect. A restricted vacuum hose would cause a hard pedal with poor braking performance.

    Answer B is correct. A leaking master cylinder would cause a low, soft pedal.

    Answer C is incorrect. A bypassing master cylinder would cause a low pedal; however, it would not cause the reservoir to be low.

    Answer D is incorrect. A stuck open vacuum check valve would cause no vacuum assist after the engine was shut off.

42. A brake line needs replaced. Technician A says that a new section of line should be installed with compression fittings. Technician B says a new section of line should be installed using an ISO or double flare. Who is correct?

    A. A only
    B. B only
    C. Both A and B
    D. Neither A nor B

    TASK A.2.4

    Answer A is incorrect. Compression fittings should never be used on hydraulic brake lines.

    Answer B is correct. Only Technician B is correct. When a steel brake line fails it should be repaired with a new section of line with either an ISO or double flare.

    Answer C is incorrect. Only Technician B is correct.

    Answer D is incorrect. Technician B is correct.

43. Which of the following is a typical input to the ABS control module?

    A. Brake light switch
    B. Pressure differential switch
    C. PRNDL switch
    D. Power steering pressure switch

    TASK F.6

    Answer A is correct. The brake light switch is a typical input to the ABS control module.

    Answer B is in correct. The ABS control module does not normally receive a signal from the pressure differential valve.

    Answer C is incorrect. The PRNDL switch is not a typical input to the ABS control module.

    Answer D is incorrect. The power steering pressure switch is not an input to the ABS control module.

Delmar, Cengage Learning ASE Test Preparation

TASK B.1

44. Technician A says that brake drag can be caused by worn drums. Technician B says that brake drag can be caused by a seized star-wheel adjuster. Who is correct?

   A. A only
   B. B only
   C. Both A and B
   D. Neither A nor B

Answer A is incorrect. Worn drums will not cause the brakes to drag. They can cause reduced stopping performance.

Answer B is incorrect. Seized star-wheel adjusters would cause the brake shoes to be out of adjustment and a low pedal complaint.

Answer C is incorrect. Neither Technician is correct.

Answer D is correct. Neither Technician is correct. Brake drag can be caused by weak return springs.

TASK A.2.1

45. A vehicle pulls left after the brake pedal is released. This could be caused by:

   A. A restricted left front brake hose.
   B. A cut master cylinder piston seal.
   C. A faulty metering valve.
   D. A restricted right front brake hose.

Answer A is correct. A restricted left front brake hose would allow higher pressure from the master cylinder to get to the caliper, but not leave the caliper. This condition would leave the left front caliper applied after the brake pedal is released.

Answer B is incorrect. A cut master cylinder piston seal would cause a lack of pressure.

Answer C is incorrect. A faulty metering valve would not normally cause a pulling condition.

Answer D is incorrect. A restricted right front hose could cause a pull to the right after the brakes are released, or a pull to the left when the brakes are first applied.

PREPARATION EXAM 2—ANSWER KEY

1. A	21. D	41. A
2. B	22. B	42. B
3. C	23. D	43. D
4. D	24. A	44. B
5. A	25. A	45. B
6. D	26. A	
7. B	27. B	
8. A	28. C	
9. D	29. C	
10. B	30. A	
11. B	31. B	
12. D	32. C	
13. C	33. B	
14. A	34. B	
15. C	35. B	
16. B	36. C	
17. A	37. B	
18. D	38. D	
19. B	39. C	
20. C	40. D	

PREPARATION EXAM 2—EXPLANATIONS

1. The brakes are dragging on a vehicle. This could be caused by:

    A. Overfilled master cylinder reservoir.
    B. Low vacuum to the booster.
    C. Bypassing master cylinder.
    D. Quick take-up valve stuck open.

    TASK A.1.1

    Answer A is correct. An overfilled master cylinder reservoir would not allow for fluid expansion. This would cause the brakes to self apply as the fluid expanded and result in brake drag.

    Answer B is incorrect. Low vacuum to the booster will cause low power assist, not dragging brakes.

    Answer C is incorrect. A bypassing master cylinder would cause a low brake pedal and poor stopping power.

    Answer D is incorrect. An open quick take-up valve would cause a low pedal complaint.

TASK C.6

2. The brake pads are worn at an angle on a vehicle with a sliding brake caliper system. Which of the following is the most likely cause?

   A. Worn caliper piston
   B. Worn caliper bushings
   C. Leaking caliper seal
   D. Leaking caliper boot

Answer A is incorrect. A worn caliper piston would most likely cause a leak.

Answer B is correct. Worn bushings can cause the caliper to cock during operation. This could cause the pads to wear at an angle.

Answer C is incorrect. A leaking caliper seal would allow air into the system and a brake fluid loss.

Answer D is incorrect. A leaking caliper boot would allow dirt to seize the piston.

TASK F.7

3. A diagnostic trouble code is set for an open wheel speed sensor on the left front sensor. The technician measures the resistance at the sensor and finds it to be within specification. Which of the following is the most likely cause for the wheel speed sensor code?

   A. Faulty ABS computer
   B. Open fuse
   C. Open wiring from the computer to the wheel speed sensor
   D. Shorted left front ABS modulator valve

Answer A is incorrect. The ABS computer could be faulty; however, an open in the sensor wiring is much more likely.

Answer B is incorrect. An open fuse would not cause this scenario. An open fuse would most likely cause the system to be inoperative.

Answer C is correct. An open in the wiring from the sensor to the computer would cause a diagnostic trouble code to be set and the sensor to have the correct resistance.

Answer D is incorrect. A shorted modulator valve would not cause a wheel speed sensor code.

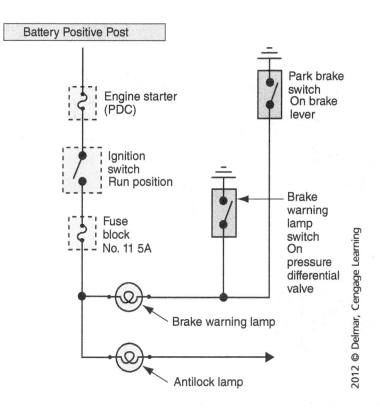

4. The brake warning lamp pictured above stays on any time the ignition switch is on. Technician A says this could be caused by a poor ground at the parking brake switch. Technician B says this could be caused by a poor ground at the brake warning lamp switch. Who is correct?

TASK E.5

A. A
B. B
C. Both A and B
D. Neither A nor B

Answer A is incorrect. A poor ground would prevent the light from coming on.

Answer B is incorrect. A poor ground would prevent the light from coming on.

Answer C is incorrect. Neither Technician is correct.

Answer D is correct. Neither Technician is correct. There is a ground present at all times to cause the light to be on all the time.

Section 6 Answer Keys and Explanations

Brakes (A5)

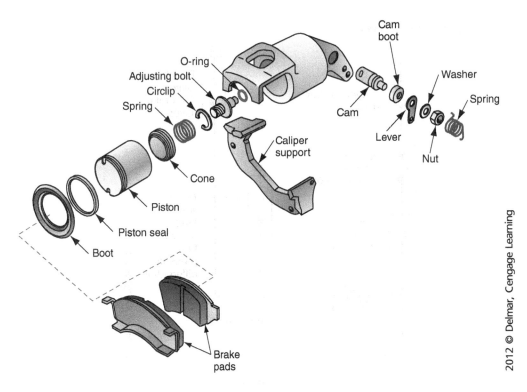

TASK C.12

5. Which of the following is true concerning the above caliper?

   A. This caliper incorporates the parking brake.
   B. This caliper can only be used on the front.
   C. This caliper can only be used with Type 5 brake fluid.
   D. This caliper must be replaced at each brake pad replacement.

   Answer A is correct. This caliper incorporates a parking brake.

   Answer B is incorrect. This caliper is used on the rear.

   Answer C is incorrect. This caliper can be used with a variety of fluids; follow the manufacturers' recommendation.

   Answer D is incorrect. The caliper is replaced only if it is faulty.

TASK A.2.4

6. Technician A says a double flare line should be replaced with an ISO flare line. Technician B says an ISO flare line should be replaced with a double flare line. Who is correct?

   A. A
   B. B
   C. Both A and B
   D. Neither A nor B

   Answer A is incorrect. A double flare line should be replaced with a double flare line.

   Answer B is incorrect. An ISO flare line should be replaced with an ISO flare line.

   Answer C is incorrect. Neither Technician is correct.

   Answer D is correct. Neither Technician is correct.

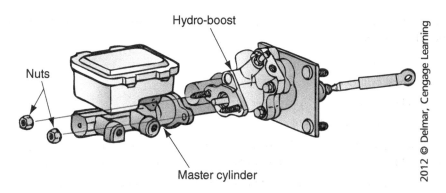

7. While testing the system shown above, the technician finds there is no assist from the booster after the engine is shut off. Which of the following is the most likely cause?

    A. A faulty back-up EH pump
    B. A faulty accumulator
    C. A bypassing master cylinder
    D. A bypassing brake booster

TASK D.4

Answer A is incorrect. The system shown has an accumulator, not a back-up pump.

Answer B is correct. The accumulator could be faulty and not storing pressure.

Answer C is incorrect. A bypassing master cylinder will cause a low, sinking pedal.

Answer D is incorrect. A bypassing brake booster would cause a problem while the engine is running as well as when the engine is stopped.

8. A customer is concerned that under heavy acceleration on dry pavement the traction control light comes on and the throttle momentarily becomes unresponsive. Which of the following is the most likely cause?

    A. A tire has lost traction.
    B. The accelerator pedal position (APP) sensor is faulty.
    C. The throttle position sensor (TPS) is faulty.
    D. A wheel speed sensor is faulty.

TASK F.3

Answer A is correct. When a tire loses traction the control module will reduce power from the engine in an attempt to regain traction. This can result in the customer sensing that the throttle is unresponsive.

Answer B is incorrect. It is unlikely that the APP sensor is faulty. It is more likely that the system is operating normally.

Answer C is incorrect. It is unlikely that the TPS is faulty. It is more likely that the system is operating normally.

Answer D is incorrect. The system is most likely acting normally.

TASK B.1

9. A vehicle has a low brake pedal. Master cylinder fluid level is correct. Which of the following is the most likely cause?

   A. Worn brake pad lining
   B. Seized wheel cylinder
   C. Seized caliper piston
   D. Brake shoes out of adjustment

   Answer A is incorrect. Since front disc brakes self adjust, worn front brake pad linings would cause low fluid level in the master cylinder reservoir. Pedal height would be acceptable.

   Answer B is incorrect. A seized wheel cylinder would cause the pedal to be firm with poor brake performance.

   Answer C is incorrect. A seized caliper piston would cause a firm pedal with poor brake performance.

   Answer D is correct. When the brake shoes are out of adjustment, the wheel cylinder must push them out further to contact the drum. The master cylinder will have to be pushed in further to accomplish this and will result in low pedal.

TASK C.7

10. The brake caliper piston boot was cut when the caliper was reassembled. Technician A says this will cause a brake fluid leak when the caliper is reinstalled. Technician B says this could cause the piston to bind due to contamination. Who is correct?

    A. A
    B. B
    C. Both A and B
    D. Neither A nor B

    Answer A is incorrect. This is not a brake fluid seal.

    Answer B is correct. Only Technician B is correct. The boot is designed to prevent contamination of the outside piston area. If the boot is ripped it could result in the piston becoming stuck.

    Answer C is incorrect. Only Technician B is correct.

    Answer D is incorrect. Technician B is correct.

TASK E.4

11. A parking brake cable is binding on a vehicle with front disc and rear drum brakes. This can result in:

    A. Dragging front brakes.
    B. Dragging rear brakes.
    C. Rusted brake drums.
    D. Rusted disc brake rotors.

    Answer A is incorrect. The parking brakes are on the rear, not the front.

    Answer B is correct. The parking brakes are on the rear; therefore, a binding cable can cause a dragging rear brake.

    Answer C is incorrect. A binding parking brake will typically cause dragging brakes. It will not cause the brake drums to rust. Brake drums rust when the service brakes do not operate, such as in the case of frozen wheel cylinders.

    Answer D is incorrect. A binding parking brake will typically cause dragging brakes. It will not cause the disc brake rotors to rust. Brake rotors rust when the service brakes do not operate, such as in the case of frozen caliper pistons.

12. Which of the following is LEAST LIKELY to be an input to the traction control system?

    A. Throttle position sensor
    B. Wheel speed sensor
    C. Engine RPM
    D. Transmission input shaft RPM

    TASK F.7

    Answer A is incorrect. The throttle position sensor is a normal input to the traction control system; this input allows the system to interpret the throttle angle.

    Answer B is incorrect. The wheel speed sensor allows the system to determine a slipping tire.

    Answer C is incorrect. The engine RPM can be an input. Some systems can reduce engine RPM to reduce wheel slippage.

    Answer D is correct. The transmission input shaft is not a typical input to the traction control system.

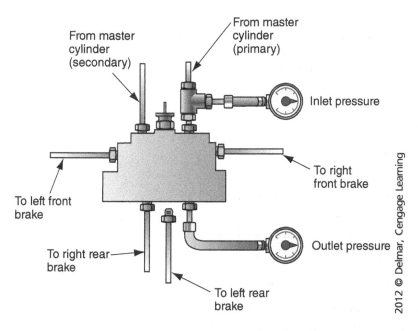

13. After a brake application, the inlet pressure gauge in the figure above drops to normal; however, the outlet pressure gauge remains higher than normal. Which of the following could be the cause?

    A. Binding brake pedal
    B. Incorrect brake pedal free stroke
    C. Restricted combination valve
    D. Weak brake shoe return springs

    TASK A.3.2

    Answer A is incorrect. A binding brake pedal would affect both gauges equally.

    Answer B is incorrect. Incorrect stroke would affect both gauges equally.

    Answer C is correct. Normally, the outlet pressure gauge would drop along with the inlet pressure gauge unless fluid is being trapped and pressure is building up. The combination valve is restricting flow. The fluid pressure is being trapped by the combination valve; the fluid is not returning to the master cylinder. The combination valve must be replaced.

    Answer D is incorrect. Worn brake shoe return springs would not cause the gauges to have different pressures.

TASK C.15

14. After disc brake pad replacement, the new pads should be burnished. This purpose of this procedure is to:

   A. Conform the pad lining to the rotor.
   B. Break the sealant off of the new linings.
   C. Remove finger prints from the rotor.
   D. Boil all the water out of the brake fluid.

   Answer A is correct. Brake pad burnishing conforms the friction material to the rotor.

   Answer B is incorrect. The purpose of burnishing is not to break the sealant off of the new linings, it is to conform the lining to the rotor.

   Answer C is incorrect. The purpose of burnishing is not to remove finger prints from the rotor, it is to conform the lining to the rotor.

   Answer D is incorrect. The purpose of burnishing is not to boil the water from the brake fluid, it is to conform the lining to the rotor.

TASK F.6

15. Technician A says some wheel speed sensors are incorporated with the wheel bearing assembly and must be replaced as a unit. Technician B says some wheel speed sensors are located in the rear axle differential. Who is correct?

   A. A
   B. B
   C. Both A and B
   D. Neither A nor B

   Answer A is incorrect. Technician B is also correct.

   Answer B is incorrect. Technician A is also correct.

   Answer C is correct. Both technicians are correct. Some sensors are located in the differential and some are incorporated into the bearing assembly.

   Answer D is incorrect. Both technicians are correct.

TASK C.4

16. Which of the following is true concerning caliper removal?

   A. The fluid must be drained prior to removal.
   B. The caliper sliding pins must be removed prior to removal.
   C. The piston must be replaced every time the caliper is removed.
   D. The boot must be replaced every time the caliper is removed.

   Answer A is incorrect. The fluid does not have to be drained to remove the caliper.

   Answer B is correct. The pins do have to be removed to remove the caliper.

   Answer C is incorrect. The piston is only replaced if it is damaged.

   Answer D is incorrect. The boot is replaced only if it is damaged.

17. The brake fluid has been contaminated with oil. Which of the following is correct?

    A. All rubber components must be replaced.
    B. Only rubber components which show contamination should be replaced.
    C. All steel lines must be replaced.
    D. Only steel lines which show contamination should be replaced.

    Answer A is correct. All rubber components must be replaced. Oil contamination will ruin all the rubber parts of a brake system.

    Answer B is incorrect. All rubber components must be replaced. The contamination of oil in the brake fluid has weakened all the rubber components, which includes hoses, o-rings and seals.

    Answer C is incorrect. The steel lines do not need to be replaced. The steel lines will need to be flushed.

    Answer D is incorrect. The steel lines do not need to be replaced. The steel lines will need to be flushed.

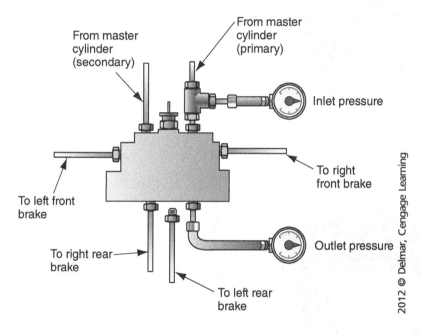

18. During a pressure test both gauges in the figure above read lower than normal. Technician A says a frozen (stuck) disc brake caliper could be the cause. Technician B says a frozen (stuck) wheel cylinder piston could be the cause. Who is correct?

    A. A
    B. B
    C. Both A and B
    D. Neither A nor B

    Answer A is incorrect. A frozen disc brake caliper will not cause low pressure.

    Answer B is incorrect. A frozen wheel cylinder piston would not cause low pressure.

    Answer C is incorrect. Neither Technician is correct.

    Answer D is correct. Neither Technician is correct. Both gauges reading low could be caused by a faulty master cylinder.

19. When driving a new race of a tapered wheel bearing into the hub, which of the following is true?
    A. Drive the race in until flush.
    B. Drive the race in until a distinct sound change occurs.
    C. Drive the race in until it is just below the surface.
    D. Drive the race in using the old wheel bearing as a driving tool.

    Answer A is incorrect. The race will not be fully seated in this position.

    Answer B is correct. This will fully seat the race in the hub. The sound will change from a ringing sound to a deep thud. Also the hub will tend to bounce when the race is fully inserted.

    Answer C is incorrect. The race will not be fully seated using this procedure.

    Answer D is incorrect. Using the old wheel bearing will damage the new race.

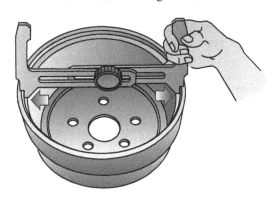

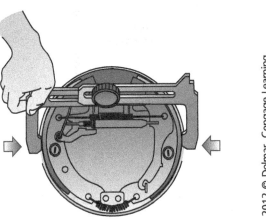

20. The tool shown above is used to:
    A. Arc the brake shoes.
    B. Compress the brake shoes.
    C. Compare the shoes to the drum.
    D. Refinish the brake drum.

    Answer A is incorrect. To arc the brake shoes involves grinding material off of the shoes. This tool will not do this.

    Answer B is incorrect. This is not a compressor.

    Answer C is correct. This tool is used to compare the drum to the shoes.

    Answer D is incorrect. This tool will not refinish the drum.

21. When the brake line is loosened at the master cylinder no fluid will drip from the loosened line; however, when the brake pedal is pulled up fluid will start to drip. Which of the following is indicated?

    A. The master cylinder should be replaced.
    B. The master cylinder should be rebuilt.
    C. The brakes may be out of adjustment.
    D. The brake pedal may be out of adjustment.

    TASK D.1

    Answer A is incorrect. The master cylinder may require further inspection. If fluid starts to flow when the pedal is lifted, the pedal is holding the piston down which prevents the free flow of brake fluid.

    Answer B is incorrect. The master cylinder may require further inspection. If moving the pedal starts the fluid flowing, the master cylinder is most likely serviceable. Adjustment of the brake pedal will be the most likely repair.

    Answer C is incorrect. This does not indicate brake adjustment.

    Answer D is correct. The brake pedal may be out of adjustment. The compensating port is being covered until the pedal is lifted.

22. When reinstalling the caliper slide pins, they should be lubricated with:

    A. High temperature wheel bearing grease.
    B. High temperature silicon grease.
    C. Lubriplate®
    D. Antiseize

    TASK C.11

    Answer A is incorrect. High temperature wheel bearing grease is not recommended for external applications; it is designed to be used in sealed areas such as wheel bearings.

    Answer B is correct. Most manufacturers recommend using a high temperature silicon grease.

    Answer C is incorrect. Lubriplate is used during engine assembly to prelube internal engine components.

    Answer D is incorrect. Antiseize is used on fasteners to help ease removal.

TASK A.3.4

23. All of the following would cause the red brake warning lamp (RBWL) to be illuminated EXCEPT:

   A. A leaking wheel cylinder.
   B. A faulty master cylinder.
   C. Broken brake hose.
   D. Low pressure to the hydraulic brake booster.

   Answer A is incorrect. A leaking wheel cylinder could cause a lack of pressure in one circuit and the pressure differential switch would turn the RBWL on.

   Answer B is incorrect. A faulty master cylinder could cause a lack of pressure in one circuit and the pressure differential switch would turn the RBWL on.

   Answer C is incorrect. A broken brake hose could cause a lack of pressure in one circuit and the pressure differential switch would turn the RBWL on.

   Answer D is correct. Low pressure to the hydraulic brake booster would not cause the red warning light to illuminate.

TASK B.10

24. Technician A says lug nuts for steel wheels should be torqued in a star pattern. Technician B says lug nuts for aluminum wheels should be torqued in a circle pattern. Who is correct?

   A. A
   B. B
   C. Both A and B
   D. Neither A nor B

   Answer A is correct. Only Technician A is correct. Lug nuts should be torqued in a star pattern regardless of wheel type.

   Answer B is incorrect. Lug nuts should be torqued in a star pattern regardless of wheel type.

   Answer C is incorrect. Only Technician A is correct.

   Answer D is incorrect. Technician A is correct.

25. Technician A says the purpose of the traction control deactivation switch is to allow wheel spin in deep snow or mud. Technician B says the purpose of the traction control deactivation switch is to improve fuel economy. Who is correct?

    A. A
    B. B
    C. Both A and B
    D. Neither A nor B

    Answer A is correct. Only Technician A is correct. The switch allows wheel spin for trying to remove a stuck vehicle from deep snow or mud.

    Answer B is incorrect. The system has no affect on fuel economy.

    Answer C is incorrect. Only Technician A is correct.

    Answer D is incorrect. Technician A is correct.

26. The brake fluid has been contaminated with oil. Technician A says all rubber hydraulic brake components must be replaced. Technician B says all the antilock braking system (ABS) wheel speed sensors must be replaced. Who is correct?

    A. A
    B. B
    C. Both A and B
    D. Neither A nor B

    Answer A is correct. Only Technician A is correct. The oil has contaminated the system and any rubber part which contacts the brake fluid must be replaced.

    Answer B is incorrect. The ABS sensors do not contact the fluid and will not be damaged by oil in the brake fluid.

    Answer C is incorrect. Only Technician A is correct.

    Answer D is incorrect. Technician A is correct.

27. When adjusting the parking brake which of the following is correct?

    A. Adjust the cable first, then the shoes.
    B. Adjust the shoes first, then the cable.
    C. Adjust the calipers first, then the shoes.
    D. Adjust the calipers first, then the cable.

    Answer A is incorrect. If the cable was adjusted first the parking brake may still not apply properly.

    Answer B is correct. The shoes should be adjusted, then the cable.

    Answer C is incorrect. Calipers do not require adjustment.

    Answer D is incorrect. Calipers do not require adjustment.

Section 6 Answer Keys and Explanations — Brakes (A5)

TASK F.7

28. The traction control lamp stays on continuously on a vehicle. The ABS light is not lit. Which of the following is the most likely cause?

    A. Wheel speed sensor
    B. ABS control module
    C. Traction control module
    D. ABS modulator valve

    Answer A is incorrect. The wheel speed sensor would also light the ABS lamp.

    Answer B is incorrect. The ABS control module would also light the ABS lamp.

    Answer C is correct. The most likely cause is a fault in the traction control module.

    Answer D is incorrect. The ABS modulator valve would most likely light the ABS lamp.

TASK A.1.3

29. Technician A says that a master cylinder pushrod that is adjusted too long could cause the brakes to drag. Technician B says that a master cylinder pushrod that is adjusted too long could result in repeat master cylinder failure. Who is right?

    A. A only
    B. B only
    C. Both A and B
    D. Neither A nor B

    Answer A is incorrect. Technician B is also correct.

    Answer B is incorrect. Technician A is also correct.

    Answer C is correct. Both technicians are correct. A master cylinder pushrod that is adjusted too long would cause the compensating port to be covered and not allow for fluid expansion. This would result in brake drag. If the pushrod is adjusted too long the master cylinder could over-travel. This over-travel condition could damage the master cylinder and cause repeat master cylinder failure.

    Answer D is incorrect. Both technicians are correct.

TASK C.5

30. The caliper slide pin bushings are worn. Which of the following is true?

    A. The bushings must be replaced.
    B. The caliper must be replaced.
    C. The bushings and the caliper piston must be replaced.
    D. The bushings must be heated to be removed.

    Answer A is correct. The bushings must be replaced.

    Answer B is incorrect. The bushings can be replaced separately from the caliper.

    Answer C is incorrect. The piston is replaced only if it is worn.

    Answer D is incorrect. The bushings do not have to be heated to be removed.

Section 6 Answer Keys and Explanations

Brakes (A5)

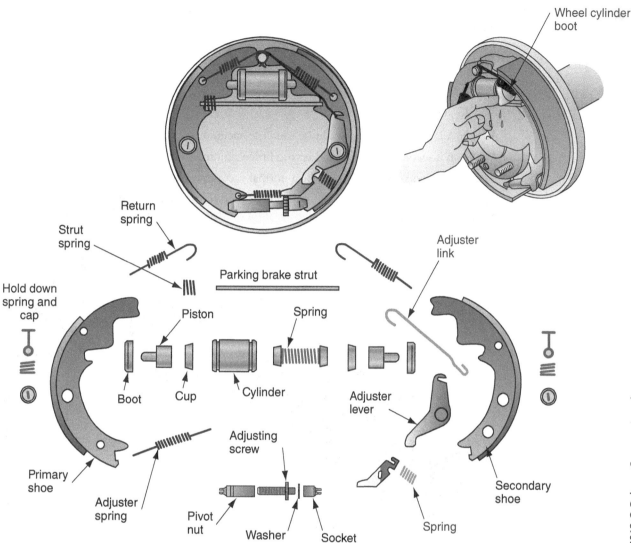

31. In the figure above, if the spring is left out during assembly, which of the following would most likely occur?

    A. The parking brake would not work.
    B. The brake would not self adjust.
    C. The service brake would only work in a rearward rotation.
    D. The service brake would only work in a forward rotation.

    Answer A is incorrect. The parking brake would operate.

    Answer B is correct. The brake would not self adjust.

    Answer C is incorrect. The service brake would work in both directions.

    Answer D is incorrect. The service brake would work in both directions.

TASK B.8

Section 6 Answer Keys and Explanations — Brakes (A5)

TASK D.2

32. Vacuum to the booster is being measured with the engine idling. Which of the following would be considered a normal reading?

   A. 5″ Hg
   B. 11″ Hg
   C. 18″ Hg
   D. 25″ Hg

   Answer A is incorrect. This would be too little vacuum.

   Answer B is incorrect. This would be too little vacuum.

   Answer C is correct. This would be a normal vacuum reading.

   Answer D is incorrect. This would be a higher than normal vacuum; however, it would not damage the booster. A gauge reading this high is extremely unlikely and would likely indicate a faulty vacuum gauge.

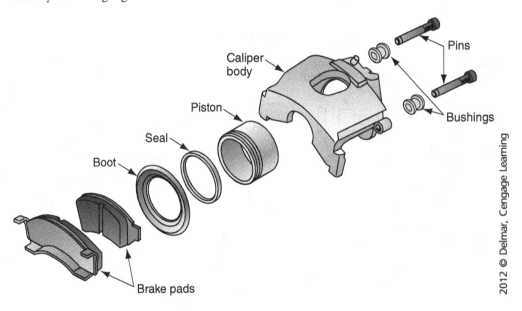

TASK C.5

33. The brake caliper shown above is a:

   A. Quad piston floating design.
   B. Single piston floating design.
   C. Dual piston fixed design.
   D. Dual piston floating design.

   Answer A is incorrect. This is a single piston floating design. A quad piston design has four pistons.

   Answer B is correct. This is a single piston floating design.

   Answer C is incorrect. This is a single piston floating design. A dual piston design has two pistons.

   Answer D is incorrect. This is a single piston floating design. A dual piston design has two pistons.

34. The flexible line connected to the front disc brake caliper is cracked. Which of the following should the technician do?

    A. Repair the section of cracked hose using a barb fitting.
    B. Replace the hose with a new one.
    C. Replace all the rubber hoses on the vehicle.
    D. Replace all the rubber and steel hoses on the vehicle.

    Answer A is incorrect. The hydraulic line should be replaced as a unit.

    Answer B is correct. The hose should be replaced.

    Answer C is incorrect. There is no need to replace all the hoses just because one is cracked.

    Answer D is incorrect. The only hose which needs to be replaced is the one which is cracked.

    TASK A.2.3

35. A vehicle equipped with front disc/rear drum brakes has a binding parking brake cable. This can result in:

    A. Difficulty removing the disc brake pads.
    B. Difficulty removing the brake drum.
    C. Difficulty installing the disc brake pads.
    D. Difficulty installing the brake shoes.

    Answer A is incorrect. This will not affect disc brake pad removal. The disc brakes are on the front and the parking brake operates the rear brakes.

    Answer B is correct. This drum may be difficult to remove, because the shoes are not fully retracted.

    Answer C is incorrect. This will not affect installing the disc brake pads. They are on the front and the parking brake operates the rear brakes.

    Answer D is incorrect. The brake drum will be difficult to remove.

    TASK E.4

36. A customer is concerned that the traction control light flashed on momentarily while driving on slick roads. The technician finds no stored diagnostic trouble codes. Which of the following is most likely the cause?

    A. A faulty wheel speed sensor
    B. A faulty brake modulator
    C. Normal operation
    D. Mismatched tires

    Answer A is incorrect. A faulty wheel speed sensor may cause the light to flash on; however, there should be a diagnostic trouble code.

    Answer B is incorrect. A faulty brake modulator valve may cause the light to flash on; however, there should be a diagnostic trouble code stored.

    Answer C is correct. When the traction control system activates it is normal for the traction control light to illuminate. This informs the operator that the system is operating.

    Answer D is incorrect. Mismatched tires may cause the light to flash on; however, a diagnostic trouble code should be stored.

    TASK F.4

TASK A.1.2

37. A step bore master cylinder has been diagnosed to be faulty. Technician A says that the brake booster must be replaced along with the master cylinder. Technician B says that the step bore master cylinder cannot be rebuilt because it cannot be honed. Who is right?

   A. A only
   B. B only
   C. Both A and B
   D. Neither A nor B

   Answer A is incorrect. A technician can replace the master cylinder and booster as an assembly; however, it is not required. They can be replaced as separate pieces.

   Answer B is correct. Only Technician B is correct. Step bore master cylinders cannot be rebuilt because the bore cannot be honed. The aluminum bore is anodized and honing would remove the hard bore coating.

   Answer C is incorrect. Only Technician B is correct.

   Answer D is incorrect. Technician B is correct.

TASK D.1

38. The hydraulic brakes on a car function correctly. However, the vacuum brake booster is not providing any brake assistance. Which of the following could be the cause?

   A. Too much vacuum to the booster
   B. Low brake fluid
   C. Incorrect brake fluid
   D. Too little vacuum to the booster

   Answer A is incorrect. Too much vacuum to the booster is extremely unlikely and would not cause low assistance from the booster.

   Answer B is incorrect. Low brake fluid would cause the brakes to be soft and not function correctly.

   Answer C is incorrect. Incorrect brake fluid could cause problems with the brakes, but it would not affect the booster's operation.

   Answer D is correct. Too little vacuum to the booster could cause little or no assistance from the booster.

TASK C.5

39. When installing the slide pins on the disc brake caliper the technician finds one hole stripped out. Which of the following is the most economical repair?

   A. The caliper must be replaced.
   B. The disc brake rotor must be replaced.
   C. There may be oversized bolts available.
   D. The steering knuckle must be replaced.

   Answer A is incorrect. The caliper is not damaged.

   Answer B is incorrect. The stripped caliper mounting hole will not affect the serviceability of the rotor. Rotors must be discarded if they are worn too thin.

   Answer C is correct. There are many models which have oversized bolts available; this is usually the easiest and most economical repair.

   Answer D is incorrect. The steering knuckle does not have to be replaced. It is not economical to replace the steering knuckle when there are more cost effective alternative repairs.

40. Technician A says DOT 3 and DOT 5 brake fluid can be mixed in a brake system without any problems. Technician B says DOT 4 and DOT 5 brake fluid can be mixed in a brake system without any problems. Who is correct?

    A. A
    B. B
    C. Both A and B
    D. Neither A nor B

    TASK A.4.4

    Answer A is incorrect. DOT 5 cannot be mixed with other brake fluids.

    Answer B is incorrect. DOT 5 cannot be mixed with other brake fluids.

    Answer C is incorrect. Neither Technician is correct.

    Answer D is correct. Neither Technician is correct. DOT 5 brake fluid cannot be mixed with any other fluid.

41. The proper procedure for adjusting tapered roller wheel bearings used on some drum brake hubs is:

    A. Tighten while spinning the hub to seat the bearings, then back off to the specified amount of end-play.
    B. Finger tighten the adjusting nut.
    C. Torque the adjusting nut to 20 lb/ft of torque and then turn the nut an additional 45 degrees.
    D. Torque the nut to 75 ft lbs with the wheels on the ground.

    TASK E.1

    Answer A is correct. Tapered roller wheel bearings should be tightened while spinning the hub to seat the bearings, and then backed off to the specified amount of end-play.

    Answer B is incorrect. This would allow the bearings to be too loose.

    Answer C is incorrect. This would cause the bearings to be too tight.

    Answer D is incorrect. This would over tighten the wheel bearings.

42. A drum brake will rotate freely forward but will bind when rotated backward. Which of the following is the most likely cause?

    A. A broken return spring on the front shoe
    B. A broken return spring on the rear shoe
    C. A binding wheel bearing
    D. An out of round drum

    TASK B.2

    Answer A is incorrect. A broken return spring on the front shoe would cause the brake to bind going forward.

    Answer B is correct. A broken return spring on the rear shoe would cause the brake to bind going backward.

    Answer C is incorrect. A binding wheel bearing would cause a binding in both directions.

    Answer D is incorrect. An out of round brake drum would not cause this concern.

Section 6 Answer Keys and Explanations — Brakes (A5)

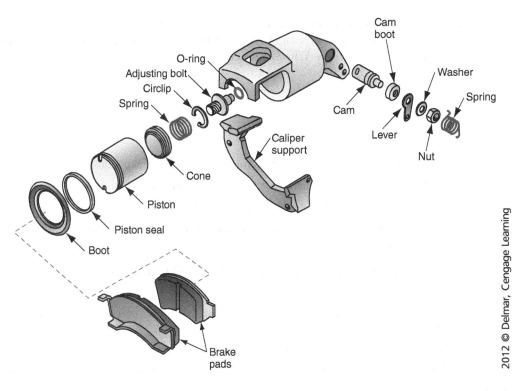

TASK E.3

43. Refer to the illustration above. The service brakes work correctly; however, the parking brake will not function. Technician A says the problem could be a torn boot. Technician B says the problem could be air in the hydraulic system. Who is correct?

A. A
B. B
C. Both A and B
D. Neither A nor B

Answer A is incorrect. A torn boot would affect both systems, because a torn boot can cause the piston to seize.

Answer B is incorrect. Air in the hydraulic system would cause the service brakes to be spongy and soft.

Answer C is incorrect. Neither Technician is correct.

Answer D is correct. Neither Technician is correct.

44. During a brake inspection on a single piston floating caliper the technician finds the inner pad worn much more than the outer. Technician A says a binding piston could be the cause. Technician B says a sticking caliper body could be the cause. Who is correct?

    A. A
    B. B
    C. Both A and B
    D. Neither A nor B

    Answer A is incorrect. If the piston was sticking the inner pad would not be pushed against the rotor; therefore, it would not wear the inner pad.

    Answer B is correct. Only Technician B is correct. This is a sliding caliper. If the caliper will not slide, the inner pad will do all the braking and the outer pad will not do any.

    Answer C is incorrect. Only Technician B is correct.

    Answer D is incorrect. Technician B is correct.

45. Brake fluid test strips can test the brake fluid for:

    A. Oil.
    B. Water.
    C. Freeze protection.
    D. Power steering fluid.

    Answer A is incorrect. Test strips do not test for the presence of oil. The presence of oil can be determined by taking a teaspoon of the brake fluid, putting a drop of water into it, and watching for the reaction.

    Answer B is correct. Test strips check for water or copper contaminated brake fluid.

    Answer C is incorrect. The test strips do not test for freeze protection.

    Answer D is incorrect. The test strip tests for water or copper.

PREPARATION EXAM 3—ANSWER KEY

1. A	21. A	41. B
2. A	22. B	42. A
3. A	23. D	43. A
4. A	24. D	44. A
5. D	25. D	45. B
6. C	26. C	
7. D	27. C	
8. D	28. C	
9. D	29. B	
10. B	30. D	
11. C	31. C	
12. A	32. D	
13. D	33. B	
14. C	34. C	
15. C	35. B	
16. C	36. D	
17. D	37. C	
18. A	38. A	
19. A	39. C	
20. B	40. C	

PREPARATION EXAM 3—EXPLANATIONS

TASK C.15

1. The goal of burnishing the brakes after a complete brake service is to:
   A. Heat the brake assembly.
   B. Cool the brake assembly.
   C. Set the wheel bearings.
   D. Stretch the brake shoes to match the drum.

   Answer A is correct. The brake assemblies are intentionally allowed to get heated during burnishing. This allows the braking materials to cure.

   Answer B is incorrect. The brakes are heated, not cooled during a burnishing procedure.

   Answer C is incorrect. Burnishing does not set the wheel bearings. The wheel bearings are set during installation.

   Answer D is incorrect. The shoes are not stretched to match the drum. The brake shoes should match the drum if correctly installed.

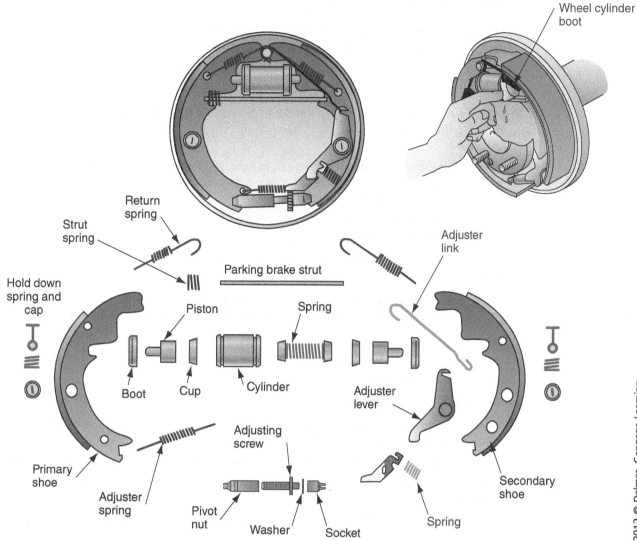

2. Refer to the figure above. The internal spring in the wheel cylinder is left out during a brake rebuild. Which of the following would be the most likely result?

    A. Air would be allowed in the brake fluid.
    B. The brake would drag.
    C. The brake would not apply.
    D. The brake would overheat.

    Answer A is correct. If the spring was missing the cups could pull away from the walls and allow air into the system.

    Answer B is incorrect. This would not cause the brake to drag; it would allow air to enter the system.

    Answer C is incorrect. This would not prevent the brake from applying; it would allow air in the system.

    Answer D is incorrect. This would not cause the brake to overheat; overheating brakes are usually caused by dragging brakes.

TASK B.7

Section 6 Answer Keys and Explanations — Brakes (A5)

3. Which of the following is true concerning master cylinder pushrod adjustment?
   A. There should be a slight gap between the pushrod and master cylinder piston.
   B. It should be adjusted until the piston is pushed in slightly.
   C. It should keep a slight pressure in the hydraulic system.
   D. It should be adjusted at each brake pad replacement.

   Answer A is correct. A slight gap should be present between the pushrod and master cylinder. This will leave the compensating port open and allow for fluid expansion.

   Answer B is incorrect. A slightly pushed in master cylinder piston would cover the compensating port which would result in brake drag. If this condition is not corrected, the master cylinder could over-travel and fail.

   Answer C is incorrect. Pressure in the hydraulic system would cause the brakes to be partially applied, which would lead to brake drag.

   Answer D is incorrect. Pushrod adjustment is not performed as part of a brake pad replacement.

4. Vacuum assist from the brake booster is present with the engine running but not present after the vehicle's engine is shut off. Which of the following could cause this condition?
   A. A faulty vacuum check valve
   B. A faulty vacuum booster hose
   C. Low manifold vacuum
   D. A leaking intake manifold

   Answer A is correct. A faulty vacuum check valve could cause no vacuum stored in the brake booster after the engine is shut off.

   Answer B is incorrect. A faulty booster hose would cause low boost with the engine running.

   Answer C is incorrect. Low manifold vacuum would cause low boost with the engine running.

   Answer D is incorrect. A leaking intake manifold would affect boost at all times.

5. Technician A says an analog wheel speed sensor generates a square wave output signal. Technician B says a digital wheel speed sensor generates an ac signal. Who is correct?
   A. A
   B. B
   C. Both A and B
   D. Neither A nor B

   Answer A is incorrect. An analog wheel speed sensor does not generate a square waveform.

   Answer B is incorrect. A digital wheel speed sensor does not generate an ac waveform.

   Answer C is incorrect. Neither technician is correct.

   Answer D is correct. Neither technician is correct. An analog wheel speed sensor generates an ac waveform, and a digital wheel speed sensor generates a dc waveform.

6. Technician A says if a vehicle has a floating rear disc brake caliper the parking brake actuation mechanism will be inside the caliper piston. Technician B says if a vehicle has a fixed rear disc brake caliper the parking brake may be inside the disc brake rotor. Who is correct?

TASK E.3

A. A
B. B
C. Both A and B
D. Neither A nor B

Answer A is incorrect. Technician B is also correct.

Answer B is incorrect. Technician A is also correct.

Answer C is correct. Both technicians are correct. A floating rear caliper can use a parking brake system which has the mechanism in the caliper piston. A fixed caliper will have the parking brake in the hat of the rotor.

Answer D is incorrect. Both technicians are correct.

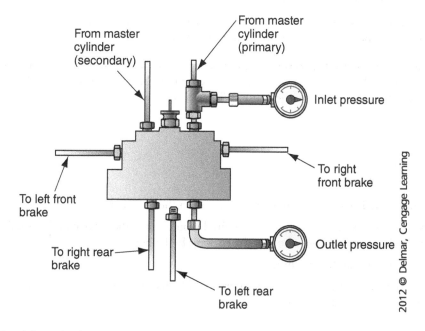

7. After a brake application, the inlet pressure gauge in the figure above drops to normal; however, the outlet pressure gauge remains higher than normal. Technician A says a faulty disc brake caliper could be the cause. Technician B says a faulty wheel cylinder could be the cause. Who is correct?

TASK A.3.2

A. A
B. B
C. Both A and B
D. Neither A nor B

Answer A is incorrect. A disc brake caliper would not cause pressure to remain high after a brake application.

Answer B is incorrect. A faulty wheel cylinder would not make the pressure remain high after a brake application.

Answer C is incorrect. Neither Technician is correct.

Answer D is correct. Neither Technician is correct. A restricted combination valve could be the cause.

TASK C.8

8. The minimum rotor thickness specification is 1.250" (31.75mm). Which of the following measurements would indicate the rotor is reusable?

   A. 1.115" (28.32mm)
   B. 1.125" (28.57mm)
   C. 1.225" (31.11mm)
   D. 1.520" (38.61mm)

   Answer A is incorrect. This measurement would mean the rotor is too thin. It is lower than the specification for minimum thickness. Thin rotors will overheat easily which results in poor brake performance.

   Answer B is incorrect. This measurement would mean the rotor is too thin. It is lower than the specification for minimum thickness. Thin rotors will develop runout during braking and this will cause the customer to feel a vibration in the brake pedal.

   Answer C is incorrect. This measurement would mean the rotor is too thin. It is lower than the specification for minimum thickness.

   Answer D is correct. This measurement is greater than specification, indicating the rotor is not worn too thin. The specification is for minimum thickness.

TASK E.3

9. When the parking brake is released the parking brake pedal fails to return to its most upward position. Technician A says a faulty master cylinder could be the cause. Technician B says air trapped in the brake fluid could be the cause. Who is correct?

   A. A
   B. B
   C. Both A and B
   D. Neither A nor B

   Answer A is incorrect. The master cylinder would not affect the return of the parking brake pedal.

   Answer B is incorrect. Air trapped in the brake fluid would make the service brake pedal feel spongy, but would not affect the return of the parking brake pedal.

   Answer C is incorrect. Neither Technician is correct.

   Answer D is correct. Neither Technician is correct. A faulty or broken return spring could cause the pedal to fail to fully return..

Section 6 Answer Keys and Explanations

Brakes (A5)

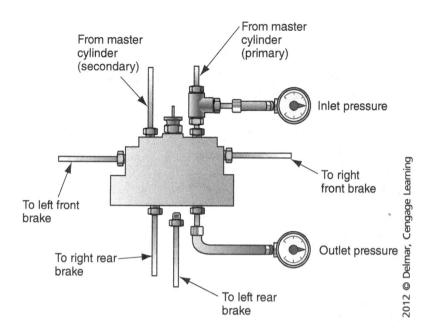

10. A vehicle with poor brake performance is being tested as shown above. Both pressure gauges are below normal. Which of the following could be the cause?

    A. A restricted combination valve
    B. A leaking cup on the primary piston
    C. A leaking cup on the secondary piston
    D. A restricted pressure differential switch

TASK A.3.1

Answer A is incorrect. A restricted combination valve would not cause both gauges to be low.

Answer B is correct. Both gauges are connected to the primary circuit; a leaking primary cup could cause only the primary pressure to be low.

Answer C is incorrect. A leaking secondary cup would not affect the primary pressure.

Answer D is incorrect. A restricted pressure differential switch would not cause the master cylinder to produce low pressure.

11. A complete brake job including installing a new master cylinder has been performed on a vehicle. After a short test drive all the brakes locked up. Which of the following is the most likely cause?

    A. Brake shoes were incorrectly installed.
    B. Brake pads were incorrectly installed.
    C. The master cylinder was incorrectly installed.
    D. The antilock braking system (ABS) sensors were not properly reset.

TASKS A.1.1, F.4

Answer A is incorrect. Incorrectly installed brake shoes would not cause all the brakes to lockup. It could cause dragging brakes.

Answer B is incorrect. Incorrectly installed brake pads would not cause all the brakes to lockup. It could cause the brakes to be noisy.

Answer C is correct. The most likely cause is that the master cylinder pushrod was not correctly adjusted.

Answer D is incorrect. Improperly resetting the ABS sensors would not cause the brakes to lockup. It could cause an ABS fault code.

Delmar, Cengage Learning ASE Test Preparation

Section 6 Answer Keys and Explanations   Brakes (A5)

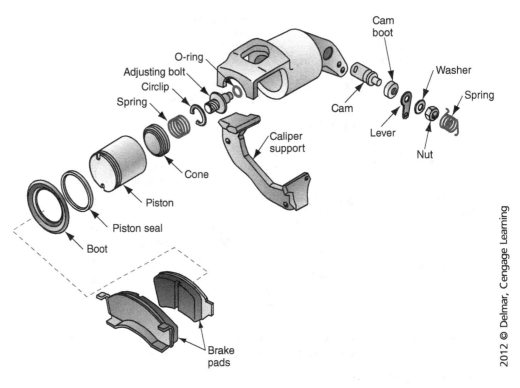

TASK C.3

12. The brake pads need to be replaced on the above system. Which of the following is true?

   A. The adjusting bolt must be retracted prior to pad installation.
   B. The adjusting bolt must be retracted prior to pad removal.
   C. The adjusting bolt must be extended prior to pad installation.
   D. The adjusting bolt must be extended prior to pad removal.

Answer A is correct. The adjusting bolt is retracted to make room for the new pads after the old pads are removed and before the new pads are installed.

Answer B is incorrect. The adjusting bolt is retracted to make room for the new pads after the old pads are removed and before the new pads are installed.

Answer C is incorrect. The adjuster is extended after the new pads are installed, as the parking brake is adjusted.

Answer D is correct. The adjusting bolt is retracted to make room for the new pads after the old pads are removed and before the new pads are installed.

13. When a customer attempts to apply the parking brake, the pedal travels its full range of motion however the parking brakes do not apply. Technician A says a faulty master cylinder could be the cause. Technician B says air trapped in the brake fluid could be the cause. Who is correct?

    A. A
    B. B
    C. Both A and B
    D. Neither A nor B

    Answer A is incorrect. The parking brake works independent of the master cylinder. The master cylinder would not cause the parking brake to fail to engage.

    Answer B is incorrect. The parking brake is applied manually. Air trapped in the brake fluid would not prevent parking brake engagement.

    Answer C is incorrect. Neither Technician is correct.

    Answer D is correct. Neither Technician is correct. This problem could be caused by the rear brakes being out of adjustment.

14. Which of the following is true concerning a non-integral ABS system?

    A. The master cylinder and booster are mounted separately.
    B. The ABS controller and master cylinder are mounted together.
    C. They can use a vacuum brake booster.
    D. They use an electro-hydraulic pump as a booster.

    Answer A is incorrect. The master cylinder and booster are mounted together as in a conventional non-ABS brake system.

    Answer B is incorrect. The ABS controller and master cylinder are separate units.

    Answer C is correct. They can use a vacuum brake booster.

    Answer D is incorrect. They do not use an electro-hydraulic pump to provide brake boost assist.

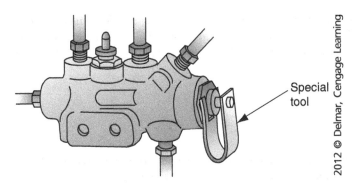

TASK A.4.2

15. The tool shown above is used on the:

   A. Master cylinder.
   B. Brake booster.
   C. Combination valve.
   D. Proportioning valve.

   Answer A is incorrect. This tool is used on the combination valve to hold the metering valve open during bleeding.

   Answer B is incorrect. This tool is used on the combination valve to hold the metering valve open during bleeding.

   Answer C is correct. This tool is used on the combination valve to hold the metering valve open during bleeding.

   Answer D is incorrect. This tool is used on the combination valve to hold the metering valve open during bleeding.

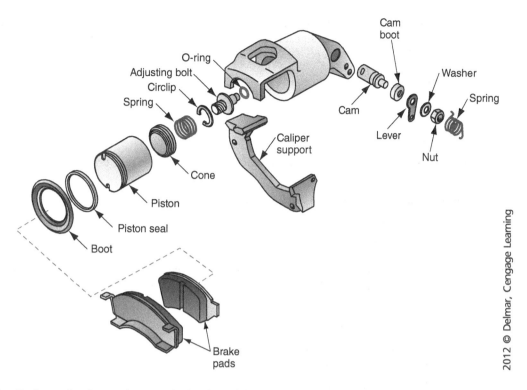

16. Refer to the figure above. Which of the following would normally be used to return the piston fully into the caliper bore?

    A. A flat tip screwdriver
    B. A cross tip (Phillips) screw driver
    C. A special tool
    D. A C-clamp

    Answer A is incorrect. A flat tip screwdriver would not be used for this job. A special tool is required.

    Answer B is incorrect. A cross tip screwdriver would not be used for this job. A special tool is required.

    Answer C is correct. A special adapter, which engages the recess of the piston, is used to turn the piston back into the caliper.

    Answer D is incorrect. A C-clamp is not used on this style caliper to fully retract the piston.

17. The screw threads of the drum brake self adjuster should be lubricated with which of the following?

    A. Oil
    B. Chassis grease
    C. Lubriplate®
    D. Antiseize

    Answer A is incorrect. Oil may contaminate the brakes and would not provide lubrication over a long period of time.

    Answer B is incorrect. Chassis grease will attract dirt and can contaminate the shoes.

    Answer C is incorrect. Lubriplate is used in engine assembly, not on the brakes.

    Answer D is correct. Antiseize is the correct lubricant.

Section 6 Answer Keys and Explanations — Brakes (A5)

TASK A.2.1

18. The right front caliper is inoperative. All other brakes operate normally. Technician A says that a collapsed brake hose could cause this. Technician B says that a faulty master cylinder could cause this. Who is correct?

    A. A only
    B. B only
    C. Both A and B
    D. Neither A nor B

    Answer A is correct. Technician A is correct. A collapsed hose may prevent pressure from reaching an individual caliper.

    Answer B is incorrect. A faulty master cylinder would affect more than just one wheel.

    Answer C is incorrect. Only Technician A is correct.

    Answer D is incorrect. Technician A is correct.

TASK E.1

19. Tapered front roller bearings were replaced during a brake job. The vehicle has come back into the shop with the wheel bearing welded to the spindle. Which of the following is the most likely cause?

    A. The bearing was overtightened.
    B. The bearing clearance was left too loose.
    C. The lug nuts were overtightened.
    D. The lug nuts were left loose.

    Answer A is correct. If the bearings were overtightened they would overheat and weld to the spindle.

    Answer B is incorrect. Loose bearings do not tend to weld themselves to the spindle. They can cause caliper piston push back and a low brake pedal.

    Answer C is incorrect. Lug nut torque would not cause the bearings to weld to the spindle. Over tightened lug nuts can cause a warped rotor and pulsating brake pedal.

    Answer D is incorrect. Loose lug nuts would not cause the bearings to weld to the spindle. Loose lug nuts can cause a damaged wheel.

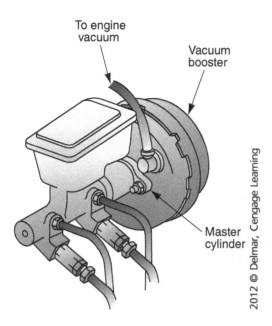

20. Technician A says the line labeled "To engine vacuum" in the figure above would connect to the intake manifold on a diesel engine. Technician B says the line labeled "To engine vacuum" would connect to the intake manifold on a gasoline engine. Who is correct?

A. A
B. B
C. Both A and B
D. Neither A nor B

TASK D.2

Answer A is incorrect. A diesel engine does not have intake manifold vacuum; this line would connect to a vacuum pump when the vehicle is equipped with a diesel engine.

Answer B is correct. Only Technician B is correct. This line does connect to the intake manifold on a gasoline engine. It is the source of vacuum for the vacuum assist unit.

Answer C is incorrect. Only Technician B is correct.

Answer D is incorrect. Technician B is correct.

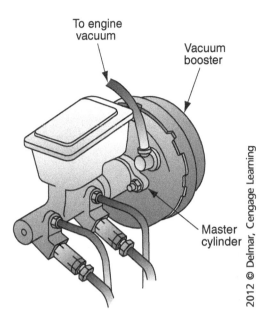

TASK A.3.2

21. In the picture shown above, the two lower brake lines are equipped with:

   A. Proportioning valves.
   B. Metering valves.
   C. Pressure differential valves.
   D. Antilock braking system (ABS) valves.

   Answer A is correct. These are proportioning valves.

   Answer B is incorrect. Metering valves are not installed in the lines as shown.

   Answer C is incorrect. A pressure differential valve will have an electrical connection.

   Answer D is incorrect. ABS valves will have electrical connections.

TASK F.4

22. Which of the following is most likely to use a single channel ABS system?

   A. Front-wheel drive
   B. Rear-wheel drive
   C. Four-wheel drive
   D. All-wheel drive

   Answer A is incorrect. A single channel ABS system is usually found on rear-wheel drive pickups and vans. Front-wheel drive vehicles usually have a three or four channel ABS system.

   Answer B is correct. A single channel ABS system is usually found on rear-wheel drive pickups and vans.

   Answer C is incorrect. Four-wheel drive vehicles tend to have four channel ABS. Some four-wheel drive vehicles will disable ABS during four-wheel drive operation.

   Answer D is incorrect. All-wheel drive vehicles tend to have four channel ABS systems.

Section 6 Answer Keys and Explanations — Brakes (A5)

23. A tapered wheel bearing has failed and spun on the spindle. The spindle now is blue/purple where the inner race sits. Besides replacing the bearing what else must be replaced?

    A. The brake pads
    B. The rotor
    C. The hub
    D. The spindle

    Answer A is incorrect. The brake pads will be replaced only if they are worn.

    Answer B is incorrect. The rotor only needs to be replaced if damaged.

    Answer C is incorrect. The hub needs to be replaced only if damaged.

    Answer D is correct. If the bearing has spun on the spindle then the spindle is worn and the new bearing will also spin.

    TASK E.1

24. On a vehicle with rear disc brakes using an integral parking brake, the service brakes work correctly; however, the parking brake will not function. Technician A says the problem could be a seized caliper piston. Technician B says the problem could be a leaking piston seal. Who is correct?

    TASK C.12

    A. A
    B. B
    C. Both A and B
    D. Neither A nor B

    Answer A is incorrect. A seized caliper piston would cause the service brakes to fail to operate.

    Answer B is incorrect. A leaking piston seal would cause the service brakes to be spongy and soft.

    Answer C is incorrect. Neither Technician is correct.

    Answer D is correct. Neither Technician is correct. A likely cause would be a stretched parking brake cable.

Section 6 Answer Keys and Explanations Brakes (A5)

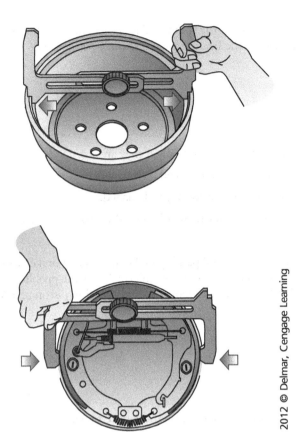

TASK B.9

25. When using the tool shown above a technician discovers that the shoes are larger than the drum. Which of the following should the technician do next?

   A. Replace the shoes.
   B. Replace the drum.
   C. Turn the drum.
   D. Turn the adjuster.

   Answer A is incorrect. If the correct shoes are installed, they do not need to be replaced, they need to be adjusted.

   Answer B is incorrect. The drum does not need to be replaced; the shoes need to be adjusted.

   Answer C is incorrect. The drum should only be turned if the finish is unacceptable. The shoes need to be adjusted to allow the drum to fit.

   Answer D is correct. The shoes need to be adjusted by turning the adjuster.

26. The technician measures the front rotors; the results are listed below:

    Minimum Rotor Thickness, Specification: .898″ (22.80mm)
    Actual Rotor Thickness, Left: .988″ (25.09mm)
    Actual Rotor Thickness, Right: .989″ (25.12mm)

    Which of the following should be done?

    A. Replace the left rotor
    B. Replace both rotors
    C. Reuse both rotors
    D. Replace the right rotor

    Answer A is incorrect. The left rotor is within specification.

    Answer B is incorrect. Both rotors are within specification.

    Answer C is correct. Both rotors are reusable.

    Answer D is incorrect. The right rotor is within specification.

27. Technician A says some parking brakes have a pedal to push to apply the parking brake. Technician B says some parking brakes have a pedal to push to release the parking brake. Who is correct?

    A. A
    B. B
    C. Both A and B
    D. Neither A nor B

    Answer A is incorrect. Technician B is also correct.

    Answer B is incorrect. Technician A is also correct.

    Answer C is correct. Both technicians are correct. Depending on the design of the parking brake, they can be used as "push to apply" or "push to release".

    Answer D is incorrect. Both technicians are correct.

28. A single channel ABS system will have ABS installed on:

    A. The front wheels only.
    B. The left side only.
    C. The rear wheels only.
    D. The right side only.

    Answer A is incorrect. A single channel ABS system will have ABS on the rear wheels only. There is not an ABS system which operates only on the front wheels.

    Answer B is incorrect. A single channel ABS system will have ABS on the rear wheels only. There is not an ABS system which only works on one side of the vehicle.

    Answer C is correct. A single channel ABS system will have ABS on the rear wheels only.

    Answer D is incorrect. A single channel ABS system will have ABS on the rear wheels only. There is not an ABS system which works on only one side of the vehicle.

TASK E.2

29. The left front wheel bearing has been replaced on a front wheel drive vehicle. Which of the following is true concerning the correct procedure to tighten the drive axle nut?

    A. The nut should be tightened then backed off one flat.
    B. The nut should be tightened to the torque specification.
    C. The nut should be adjusted to allow the wheel bearing to have 0. 0010″–0.0050″ (0.0254mm–0.127mm) end-play.
    D. The self locking nut should be reused.

    Answer A is incorrect. This would allow the drive axle to be loose.

    Answer B is correct. This is the correct procedure.

    Answer C is incorrect. This would allow the drive axle to be loose. The bearings should not have end-play on this style of bearing.

    Answer D is incorrect. Self-locking nuts should not be reused, they should be replaced.

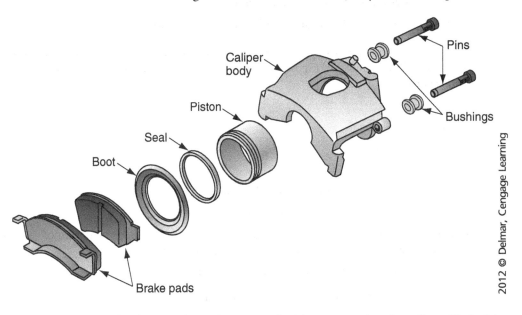

TASK C.4

30. Technician A says the above pads can be removed without removing the caliper. Technician B says the above pads can be removed without removing the pins. Who is correct?

    A. A
    B. B
    C. Both A and B
    D. Neither A nor B

    Answer A is incorrect. The caliper must be removed to remove the pads.

    Answer B is incorrect. The pins must be removed so the caliper can be removed. The caliper must be removed to remove the pads.

    Answer C is incorrect. Neither Technician is correct.

    Answer D is correct. Neither Technician is correct.

Section 6 Answer Keys and Explanations

Brakes (A5)

31. Technician A says that a leaking wheel cylinder can be identified by looking behind the boot. Technician B says that some wheel cylinders cannot be honed. Who is correct?

    A. A only
    B. B only
    C. Both A and B
    D. Neither A nor B

    TASK B.7

    Answer A is incorrect. Technician B is also correct.

    Answer B is incorrect. Technician A is also correct.

    Answer C is correct. Both technicians are correct. Leaking wheel cylinders can be detected by lifting the boot and looking behind it for brake fluid. Some wheel cylinders have hardened bores and should not be honed.

    Answer D is incorrect. Both technicians are correct.

32. Which of the following is true concerning an integral ABS system?

    A. The master cylinder and booster are separately mounted.
    B. The ABS controller and master cylinder are separately mounted.
    C. They use a vacuum brake booster.
    D. They use an electro-hydraulic pump.

    TASK F.4

    Answer A is incorrect. The master cylinder and booster are mounted together on an integral system.

    Answer B is incorrect. The ABS controller and master cylinder are mounted together on an integral system.

    Answer C is incorrect. A vacuum booster is not used on an integral ABS system.

    Answer D is correct. An integral system will use an electro-hydraulic pump to provide brake boost assist.

33. The left front disc brake caliper is dragging. All the other brakes operate normally. Which of the following could be the cause?

    A. Damaged right front brake line
    B. Damaged left front brake line
    C. Faulty master cylinder primary piston seal
    D. Faulty master cylinder secondary piston seal

    TASK A.2.1

    Answer A is incorrect. The right front brake line would not cause the left front brake to drag.

    Answer B is correct. A damaged hose can act as a check valve and cause the fluid to fail to return to the master cylinder and the brake to drag.

    Answer C is incorrect. A faulty primary piston seal would cause an internal leak in the master cylinder.

    Answer D is incorrect. A faulty secondary piston seal would cause an internal leak in the master cylinder.

Section 6 Answer Keys and Explanations — Brakes (A5)

TASK A.1.4

34. The brake pedal on a vehicle will slowly drop while the vehicle is sitting at a stop light. Which of the following could be the cause?

   A. A restricted vacuum hose
   B. Externally leaking master cylinder
   C. A bypassing master cylinder
   D. A stuck open vacuum check valve

   Answer A is incorrect. A restricted vacuum hose would cause a hard pedal with poor braking performance.

   Answer B is incorrect. A leaking master cylinder would cause a low, soft pedal, but would not cause the pedal to slowly drop while holding constant pressure.

   Answer C is correct. A bypassing master cylinder could cause the pedal to fade away while a constant pressure is applied.

   Answer D is incorrect. A stuck open vacuum check valve would cause no vacuum assist after the engine was shut off.

TASK C.3

35. The brake pads are worn at an angle on a vehicle equipped with rear disc brakes. Which of the following is the most likely cause?

   A. A stuck piston
   B. A bent caliper support
   C. A stuck adjuster
   D. A bent parking brake lever

   Answer A is incorrect. A stuck piston could cause the brake to drag or be inoperative, but would not cause the shoes to wear at an angle

   Answer B is correct. A bent support would cause the pads to contact the rotor at an angle and result in pads which are worn at an angle.

   Answer C is incorrect. A stuck adjuster may cause the parking brake to fail to adjust; however, it would not cause pads to be worn at an angle.

   Answer D is incorrect. A bent lever would not affect the angle at which the pads contact the rotor.

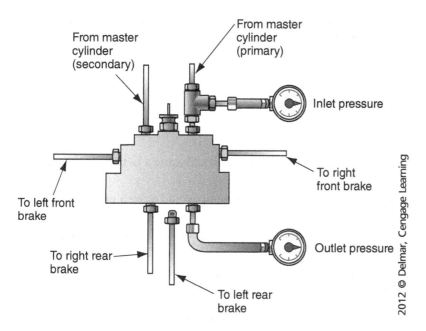

36. During a pressure test both gauges in the figure above read lower than normal. Technician A says a swollen hose could be the cause. Technician B says a restricted hose could be the cause. Who is correct?

   A. A
   B. B
   C. Both A and B
   D. Neither A nor B

TASK A.2.1

Answer A is incorrect. The hoses are after the gauges; therefore, a swollen hose could not be the cause for the low gauge readings.

Answer B is incorrect. There are no hoses between the master cylinder and the combination valve; therefore, a restricted hose would not cause this pressure decrease.

Answer C is incorrect. Neither Technician is correct.

Answer D is correct. Neither Technician is correct. There are no rubber hoses between the master cylinder and the combination valve, so this effect could only be caused by a master cylinder not producing enough pressure or by a kinked steel line prior to the gauges.

37. Technician A says an analog wheel speed sensor can be checked with a DMM set on AC volts. Technician B says an analog wheel speed sensor can be checked with an ohmmeter. Who is correct?

   A. A
   B. B
   C. Both A and B
   D. Neither A nor B

TASK F.7

Answer A is incorrect. Technician B is also correct.

Answer B is incorrect. Technician A is also correct.

Answer C is correct. Both Technicians are correct. Analog wheel speed sensors generate an AC voltage and can be measured for AC voltage output, and the sensor coil can be checked for correct resistance using an ohmmeter.

Answer D is incorrect. Both technicians are correct.

Section 6 Answer Keys and Explanations — Brakes (A5)

TASK D.4

38. A hydro-boost system is being checked on a vehicle. The power steering system functions normally; however, the brakes have very little power assist. Which of the following could be the cause?

    A. Faulty hydro-boost unit
    B. Faulty power steering pump
    C. Leaking steering gear
    D. Sticking tensioner

    Answer A is correct. A faulty hydro-boost unit would cause a lack of assist, with the power steering system working normally.

    Answer B is incorrect. A faulty power steering pump would cause a problem with the power steering system as well.

    Answer C is incorrect. A leaking steering gear would affect the steering system as well as the brake power assist system.

    Answer D is incorrect. A sticking tensioner could cause the drive belt to slip; however, this would also affect the power steering system.

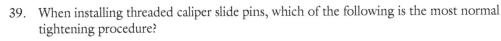
TASK C.11

39. When installing threaded caliper slide pins, which of the following is the most normal tightening procedure?

    A. Tighten to 100 ft lbs, then tighten to 250 ft lbs.
    B. Tighten to 100 ft lbs then back off 1/2 turn.
    C. Tighten to 25 ft lbs.
    D. Tighten to 50 ft lbs and back off one turn.

    Answer A is incorrect. Two hundred fifty ft lbs would be too tight for this size fastener.

    Answer B is incorrect. The pins would not be backed off; this would allow them to be loose.

    Answer C is correct. Twenty-five ft lbs would be a normal torque.

    Answer D is incorrect. The pins would not be backed off; this would allow them to be loose.

TASK A.1.1

40. The brake pedal of a vehicle slowly falls to the floor. No leaks are found and the reservoir is full. Technician A says that the cup seals in the master cylinder may be bypassing. Technician B says that a corroded internal master cylinder bore could be the cause. Who is correct?

    A. A only
    B. B only
    C. Both A and B
    D. Neither A nor B

    Answer A is incorrect. Technician B is also correct.

    Answer B is incorrect. Technician A is also correct.

    Answer C is correct. Both technicians are correct. An internally bypassing master cylinder could cause a low pedal complaint. A corroded master cylinder bore could cut the cup seals of the master cylinder pistons and cause a bypassing condition. This bypassing condition could cause a low pedal complaint.

    Answer D is incorrect. Both technicians are correct.

Section 6 Answer Keys and Explanations

Brakes (A5)

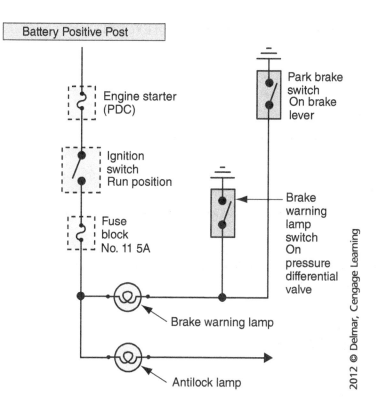

41. The brake warning lamp pictured above will come on when the service brakes are applied. Which of the following could be the cause?

    A. A misadjusted parking brake switch
    B. Air in the brake fluid
    C. Worn parking brakes
    D. Worn brake drums

TASK E.5

Answer A is incorrect. A misadjusted parking brake switch would not have any connection to the light coming on when the service brakes are applied.

Answer B is correct. Air in the fluid would cause the pressure differential switch to turn the light on when the brake pedal is depressed. This is because the air would cause a difference in pressure in the two systems.

Answer C is incorrect. Worn parking brakes would not cause the light to come on.

Answer D is incorrect. Worn brake drums would not cause the light to come on.

Section 6 Answer Keys and Explanations

Brakes (A5)

TASK B.7

42. All of the following could cause the wheel cylinder to leak EXCEPT:
    A. Brake pedal adjustment.
    B. Lack of fluid maintenance.
    C. Torn dust boots.
    D. Recent shoe replacement.

Answer A is correct. Out of adjustment brake pedal may cause the brakes to drag; however, it would not cause the wheel cylinder to leak.

Answer B is incorrect. Lack of fluid maintenance could cause rust to form inside the wheel cylinder which could cut the seals and cause a leak.

Answer C is incorrect. Torn dust boots could allow dirt and debris to cut the wheel cylinder seals and cause a leak.

Answer D is incorrect. When new brake shoes are installed, the wheel cylinder pistons are pushed back into the bore. Corrosion or debris that may be in the wheel cylinder could cut the wheel cylinder seals and cause a leak.

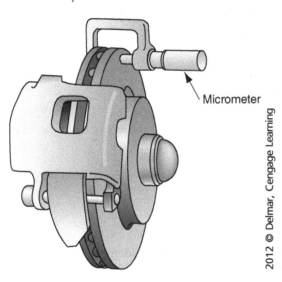

TASK C.8

43. Technician A says rotor thickness is being measured in the figure. Technician B says rotor runout is being measured. Who is correct?
    A. A
    B. B
    C. Both A and B
    D. Neither A nor B

Answer A is correct. Only Technician A is correct. Rotor thickness is being measured.

Answer B is incorrect. Runout is measured with a dial indicator.

Answer C is incorrect. Only Technician A is correct.

Answer D is incorrect. Technician A is correct.

44. A brake system which uses a vacuum brake booster has poor braking performance and a hard, high brake pedal with the engine running. Which of the following could be the cause?

    A. A restricted vacuum hose
    B. A leaking master cylinder
    C. A bypassing master cylinder
    D. A stuck open vacuum check valve.

    TASK D.2

    Answer A is correct. A restricted vacuum hose would cause a hard pedal with poor braking performance.

    Answer B is incorrect. A leaking master cylinder would cause a low, soft pedal.

    Answer C is incorrect. A bypassing master cylinder would cause a low pedal.

    Answer D is incorrect. A stuck open vacuum check valve would cause no vacuum assist after the engine was shut off.

45. When bench bleeding the master cylinder which statement is true?

    A. The piston should be stroked until bubbles start to flow.
    B. The piston should be stroked until bubbles stop flowing.
    C. Mineral spirits should be used.
    D. Mineral oil should be used.

    TASK A.1.6

    Answer A is incorrect. If the technician quit bench bleeding when bubbles started to flow all the air would not be removed.

    Answer B is correct. The wooden dowel should be used to stroke the master cylinder until the bubbles stop flowing.

    Answer C is incorrect. Mineral spirits should never be used on brake the parts which will contact brake fluid.

    Answer D is incorrect. Mineral oil should never be used on brake the parts which will contact brake fluid.

Section 6 Answer Keys and Explanations   Brakes (A5)

PREPARATION EXAM 4—ANSWER KEY

1. B
2. A
3. A
4. B
5. C
6. D
7. B
8. C
9. D
10. C
11. B
12. B
13. C
14. D
15. C
16. C
17. D
18. D
19. C
20. D
21. C
22. C
23. B
24. B
25. D
26. D
27. B
28. D
29. D
30. A
31. C
32. C
33. A
34. B
35. C
36. B
37. D
38. B
39. D
40. A
41. D
42. A
43. B
44. D
45. D

PREPARATION EXAM 4—EXPLANATIONS

TASK F.4

1. Four channel ABS sensors will have how many ABS wheel speed sensors?

   A. Two
   B. Four
   C. Six
   D. Eight

   Answer A is incorrect. A rear-wheel antilock system may have one or two ABS wheel speed sensors.

   Answer B is correct. Four channel ABS systems will have four sensors, one for each wheel.

   Answer C is incorrect. No current production system has six wheel speed sensors.

   Answer D is incorrect. No current production system has two sensors per wheel.

TASK A.3.2

2. The metering valve:

   A. Limits front brake application until the rear brakes apply.
   B. Limits rear brake application until the front brakes apply.
   C. Limits front brake application based on vehicle height.
   D. Limits rear brake application based on vehicle height.

Page 172   Delmar, Cengage Learning ASE Test Preparation

Answer A is correct. The metering valve allows the rear brakes to apply before the front brakes.

Answer B is incorrect. The metering valve allows the rear brakes to apply first.

Answer C is incorrect. The metering valve is not height sensitive.

Answer D is incorrect. The metering valve is not height sensitive.

3. A wheel cylinder has been honed. It should be cleaned with:

   A. Alcohol.
   B. Mineral spirits.
   C. Penetrating solvent.
   D. Rust preventative.

TASK B.7

Answer A is correct. Alcohol will flush the bore without leaving an oil residue.

Answer B is incorrect. Mineral spirits would leave an oil residue.

Answer C is incorrect. Penetrating solvent leaves an oil residue.

Answer D is incorrect. Rust preventative leaves an oil residue.

4. A vehicle is equipped with a disc/drum brake system. When the customer pulls the parking brake lever the lever travels its full range of motion, but the parking brake does not apply. Technician A says the disc brakes could be out of adjustment. Technician B says the drum brakes could be out of adjustment. Who is correct?

   A. A
   B. B
   C. Both A and B
   D. Neither A nor B

TASK E.4

Answer A is incorrect. The disc brakes are on the front. The parking brake is on the rear.

Answer B is correct. Only Technician B is correct. Out of adjustment rear drum brakes could be the cause.

Answer C is incorrect. Only Technician B is correct.

Answer D is incorrect. Technician B is correct.

5. A height sensing proportioning valve would be located on:

   A. The left front.
   B. The right front.
   C. The rear in the center.
   D. Only on all wheel drive vehicles.

TASK A.3.3

Answer A is incorrect. The valve is not located in the left front.

Answer B is incorrect. The valve is not located in the right front.

Answer C is correct. The height sensing valve is located in the rear in the center.

Answer D is incorrect. The vehicle does not have to be an all wheel drive to use a height sensing proportioning valve.

Section 6 Answer Keys and Explanations

Brakes (A5)

TASK F.2

6. The customer feels a rapid pedal pulsation under normal braking conditions. This could be caused by all of the following EXCEPT:

   A. Different size tires.
   B. A damaged reluctor wheel.
   C. Excessive reluctor to sensor gap.
   D. Open circuit to the speed sensor.

   Answer A is incorrect. Different size tires could cause false engagement of the ABS system since the controller will see different wheel speeds from the different sized tire(s).

   Answer B is incorrect. A damaged reluctor wheel could cause the controller to see a brief change in wheel speed. This could be interpreted as wheel lockup and the ABS would engage.

   Answer C is incorrect. Excessive reluctor to sensor gap would cause the signal to differ from the other sensors. This could be interpreted as wheel lockup and the ABS would engage.

   Answer D is correct. An open in the wheel speed sensor circuit would cause a DTC to be set. This would not cause ABS activation under normal braking conditions.

TASK C.8

7. The technician measures the front rotors; the results are listed below:

   Minimum Rotor Thickness, Specification: .898" (22.80mm)
   Actual Rotor Thickness, Left: .889" (22.58mm)
   Actual Rotor Thickness, Right: .809" (25.12mm)

   Which of the following should be done?

   A. Replace the left rotor.
   B. Replace both rotors.
   C. Reuse both rotors.
   D. Replace the right rotor.

   Answer A is incorrect. The right rotor is also out of specification.

   Answer B is correct. Both rotors are out of specification.

   Answer C is incorrect. Neither rotor is within specification.

   Answer D is incorrect. The left rotor is also out of specification.

TASK B.8

8. Technician A says the primary shoe goes toward the front of the vehicle. Technician B says the secondary shoe goes toward the rear of the vehicle. Who is right?

   A. A
   B. B
   C. Both A and B
   D. Neither A nor B

   Answer A is incorrect. Technician B is also correct.

   Answer B is incorrect. Technician A is also correct.

   Answer C is correct. Both technicians are correct. This is the correct way to install brake shoes.

   Answer D is incorrect. Both technicians are correct.

9. Which of the following would most commonly be used as the drive axle retaining nut on a front wheel drive vehicle?

    A. Self-locking nylon insert nut
    B. Plain hex nut without a locking feature
    C. Double nuts with a lock between them
    D. Self-locking crimp style nut

    TASK E.1

    Answer A is incorrect. A nylon insert nut will sometimes be found on vehicle trim pieces, but not on the drive axle.

    Answer B is incorrect. A plain hex nut is not normally used. A plain hex nut without a locking feature will not provide the safety margin needed to ensure the drive axle does not come loose during operation.

    Answer C is incorrect. Double nuts with a lock between them are sometimes used on a full floating rear drive axle bearing set.

    Answer D is correct. A self-locking crimp style nut is normally used. This nut effectively retains the torque on the drive axle.

10. The rotor thickness specification is 1.250″ (31.75mm). In three places the rotor measures 1.350″ (34.29mm). In a fourth place the rotor measures 1.248″ (31.70mm). Which of the following is true?

    A. The rotor should be turned to the smallest measurement and reused.
    B. The rotor should be reused as is.
    C. The rotor should be replaced.
    D. The measurement was performed incorrectly.

    TASK C.8

    Answer A is incorrect. The rotor is too thin and must be replaced.

    Answer B is incorrect. The rotor is too thin and must be replaced.

    Answer C is correct. This is correct. The rotor must be replaced.

    Answer D is correct. The measurement was performed correctly. When the technician finds a thickness smaller than specification the rotor must be discarded.

11. Technician A says the same tool is used to make a double flare or an ISO flare. Technician B says the same tool is used to make a double flare and a single flare. Who is correct?

    A. A
    B. B
    C. Both A and B
    D. Neither A nor B

    TASK A.2.4

    Answer A is incorrect. The double flare and the ISO flare require two different tools.

    Answer B is correct. Only Technician B is correct. The same tool is used to make a double flare and a single flare.

    Answer C is incorrect. Only Technician B is correct.

    Answer D is incorrect. Technician B is correct.

12. Which of following statements is true concerning the brake warning lights?
    A. The master cylinder can cause the amber light to illuminate.
    B. The master cylinder can cause the red light to illuminate.
    C. If the amber light is illuminated the vehicle must be pulled to the side of the road and towed.
    D. If the amber light is illuminated there is a possible hydraulic imbalance.

    Answer A is incorrect. A master cylinder failure can illuminate the red light.

    Answer B is correct. A faulty master cylinder can cause a hydraulic imbalance and cause the red light to illuminate.

    Answer C is incorrect. The amber light is the ABS light; the vehicle can be driven to a shop to be repaired with the amber light illuminated.

    Answer D is incorrect. The red light illuminates when there is a hydraulic imbalance or the parking brake is engaged.

13. The technician is preparing to do a brake job on a vehicle equipped with a non-integral ABS system. Which of the following should the technician do?
    A. Pump the pedal 25–40 times.
    B. Support the vehicle on a jack.
    C. Support the vehicle on jack stands.
    D. Release the vacuum from the brake booster.

    Answer A is incorrect. The brake pedal does not need to be pumped on a non-integral system.

    Answer B is incorrect. Working on a vehicle while it is supported only by the jack is considered an unsafe practice. If the seal leaks on the jack, the vehicle will lower.

    Answer C is correct. The vehicle must be supported on jack stands.

    Answer D is incorrect. There is no need to bleed the vacuum from the booster.

14. The technician has measured the rotor in 12 locations. Which variation in the readings would indicate a failed rotor?
    A. 0.0001" (0.00254mm)
    B. 0.0005" (0.0127mm)
    C. 0.001" (0.0254mm)
    D. 0.005" (0.127mm)

    Answer A is incorrect. This measurement passes; 0.0001 is smaller than the normal maximum specification of 0.001 (0.0254mm).

    Answer B is incorrect. This measurement passes; 0.0005 is smaller than the normal maximum specification of 0.001 (0.0254mm).

    Answer C is incorrect. This measurement passes; 0.001 is the normal maximum specification.

    Answer D is correct. This measurement fails; 0.005 is larger than the normal maximum specification of 0.001 (0.0254mm).

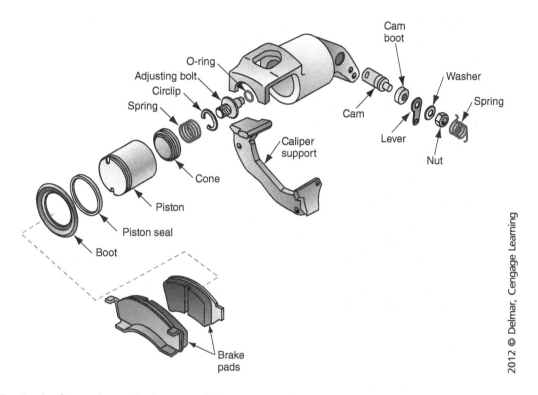

15. In the figure above, the lever would be connected to:

    A. The master cylinder.
    B. The sway bar.
    C. The parking brake cable.
    D. The brake pedal.

    TASK C.3

    Answer A is incorrect. The master cylinder is connected through the hydraulic brake line.

    Answer B is incorrect. The sway bar does not connect to the brake caliper.

    Answer C is correct. The lever connects to the parking brake cable.

    Answer D is incorrect. The brake pedal connects to the master cylinder.

16. The flexible rubber brake hose connecting to the caliper is being replaced. Which of the following should also be replaced?

    A. The caliper
    B. The banjo bolt
    C. The sealing washers
    D. The steel brake line

    TASK A.2.3

    Answer A is incorrect. The caliper does not need to be replaced with the hose.

    Answer B is incorrect. The banjo bolt does not have to be replaced with the hose.

    Answer C is correct. The sealing washers should be replaced any time they are removed.

    Answer D is incorrect. The steel brake line does not need to be replaced with the flexible hose.

17. An ABS hydraulic control unit (HCU) makes a clicking noise for approximately 5 seconds after starting the vehicle. Which of the following is indicated?

   A. Faulty ABS module
   B. Faulty ABS hydraulic unit
   C. The ABS electronic control module (ECM) is bleeding the brakes.
   D. The ABS ECM is performing a self-test.

   Answer A is incorrect. The clicking noise is a self-test being performed; a faulty ABS module would not click.

   Answer B is incorrect. The clicking noise is a self-test being performed. An ABS hydraulic unit with faulty solenoids would not click.

   Answer C is incorrect. The ABS ECM will pulse the system to help bleed the brakes when commanded to do so by the technician using a scan tool.

   Answer D is correct. This is a normal condition. A self-test is being performed

18. Technician A says the primary shoe has the long pad. Technician B says the secondary shoe is the short pad. Who is correct?

   A. A
   B. B
   C. Both A and B
   D. Neither A nor B

   Answer A is incorrect. The primary shoe does not have the long pad.

   Answer B is incorrect. The secondary shoe is not the short pad.

   Answer C is incorrect. Neither Technician is correct.

   Answer D is correct. Neither Technician is correct. The primary shoe is the short shoe, and the secondary shoe is the long shoe.

19. Which is true concerning measuring a rotor for thickness variation?

   A. The rotor should be hot when this measurement is taken.
   B. The wheel bearing should be tightened when the measurement is taken.
   C. The measurement should be taken in several locations around the rotor.
   D. The caliper must be removed before this measurement can be taken.

   Answer A is incorrect. The rotor does not need to be hot to take this measurement.

   Answer B is incorrect. Wheel bearing adjustment does not affect this measurement.

   Answer C is correct. The measurement should be taken at several locations to make sure that the rotor exceeds minimum thickness specifications in all locations.

   Answer D is incorrect. The caliper can stay on while this measurement is taken, as long as it is not in the way.

20. The master cylinder cover is removed and the steering wheel is rocked back and forth. The fluid in the master cylinder moves as the steering wheel is turned. Which of the following is the most likely cause?

    A. Warped brake rotor
    B. Worn caliper piston
    C. Seized sliding caliper
    D. Loose wheel bearings

    TASK E.1

    Answer A is incorrect. A warped rotor would not cause the fluid to be pushed back into the master cylinder during this test.

    Answer B is incorrect. A worn caliper piston would not cause the fluid to move.

    Answer C is incorrect. A seized caliper would not cause the fluid to move.

    Answer D is correct. A loose wheel bearing would cause relative motion between the rotor and caliper and cause the brake fluid to be pushed back into the reservoir.

    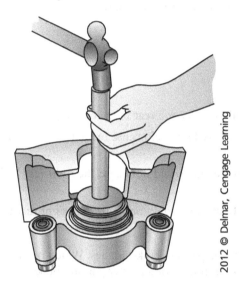

21. What is being installed in the figure above?

    A. The piston
    B. The piston seal
    C. The boot
    D. The square cut o-ring

    TASK C.11

    Answer A is incorrect. The piston would not forced in with a driving tool.

    Answer B is incorrect. The seal is installed with with a pick or by hand.

    Answer C is correct. The boot is being installed.

    Answer D is incorrect. The square cut o-ring would be installed with a pick or by hand.

TASK F.4

22. Technician A says ABS trouble codes can usually be retrieved using a scan tool. Technician B says ABS codes can usually be cleared using a scan tool. Who is correct?

   A. A
   B. B
   C. Both A and B
   D. Neither A nor B

   Answer A is incorrect. Technician B is also correct.

   Answer B is incorrect. Technician A is also correct.

   Answer C is correct. Both technicians are correct. Nearly all newer systems can retrieve and clear the ABS trouble codes using a scan tool.

   Answer D is incorrect. Both technicians are correct.

TASK A.2.1

23. The right rear brake is dragging on a vehicle with independent rear suspension. All other brakes operate normally. Which of the following is could be the cause?

   A. Damaged right front brake line
   B. Damaged right rear brake line
   C. Damaged left rear brake line
   D. Damaged left front brake line

   Answer A is incorrect. A front brake line would not cause a rear brake to drag.

   Answer B is correct. Damage to a right rear brake line could cause brake pressure to be trapped and the brake to drag.

   Answer C is incorrect. A damaged left rear brake line would not cause the right rear brake to drag.

   Answer D is incorrect. A damaged left front brake line would not cause the right rear brake to drag.

TASK E.4

24. The technician is counting the "clicks" before the parking brake applies. Which of the following is generally considered the maximum number for a proper parking brake adjustment?

   A. 5
   B. 10
   C. 15
   D. 20

   Answer A is incorrect. This would be too tight of an adjustment and could result in a dragging brake.

   Answer B is correct. Ten clicks is generally considered the maximum number of clicks.

   Answer C is incorrect. This would be too loose.

   Answer D is incorrect. This would be too loose.

25. What is the proper brake bleeding sequence?

    A. RR, LF, LR, RF
    B. LR, RF, LF, RR
    C. LR, LF, RR, RF
    D. RR, LR, RF, LF

    Answer A is incorrect. The bleeding sequence should start at the wheel furthest from the master cylinder, then progress to the wheel which is the next furthest away.

    Answer B is incorrect. The bleeding sequence should start at the wheel furthest from the master cylinder. If the technician starts at a different wheel it is possible to leave air trapped in the lines.

    Answer C is incorrect. The bleeding sequence should start at the wheel furthest from the master cylinder.

    Answer D is correct. The bleeding sequence should start at the wheel furthest from the master cylinder, then proceed to the next furthest wheel from the master cylinder.

26. A pressure gauge is installed in the bleeder screw port of a disc brake caliper, the brake pedal is depressed, and the gauge pressure indicates 750 psi. The pedal is held steady and the engine is started; the gauge pressure does not change. Which of the following is indicated?

    A. This is a normal condition.
    B. The combination valve is leaking.
    C. The combination valve is restricted.
    D. The brake booster is not functioning.

    Answer A is incorrect. The pressure should rise when the engine is started; this is not a normal condition.

    Answer B is incorrect. A leaking combination valve would not allow the gauge to reach 750 psi prior to starting the engine.

    Answer C is incorrect. A restricted combination valve would not allow the gauge to reach 750 psi prior to starting the engine.

    Answer D is correct. The pressure should rise when the engine is started; the booster is not functioning.

27. Which of the following would best describe the burnishing procedure?

    A. Five complete stops from 30 MPH
    B. Twenty slow-downs from 50 MPH to 30 MPH
    C. Five slow-downs from 50 MPH to 30 MPH
    D. Fifty slow-downs from 50 MPH to 30 MPH

    Answer A is incorrect. This is not enough to heat the brakes thoroughly.

    Answer B is correct. Twenty slow-downs from 50 MPH to 30 MPH would be an acceptable break-in method.

    Answer C is incorrect. This is not enough to heat the brakes.

    Answer D is incorrect. This would overheat the brakes.

Section 6 Answer Keys and Explanations — Brakes (A5)

TASK F.7

28. With the vehicle supported on a frame hoist and the ignition on, the technician spins one front wheel by hand. The ABS lamp illuminates. This indicates:

    A. A faulty front wheel speed sensor.
    B. A faulty rear wheel speed sensor.
    C. A faulty ABS control unit.
    D. Normal operation.

    Answer A is incorrect. The lamp illuminated because the control unit sensed that one wheel speed was generating a wheel speed signal while the others were stationary. The speed sensor is working correctly.

    Answer B is incorrect. There is no indication that there is a problem with the rear speed sensors.

    Answer C is incorrect. The ABS control unit is functioning as it should.

    Answer D is correct. This is a quick test to see if the wheel speed sensor is sending a speed signal.

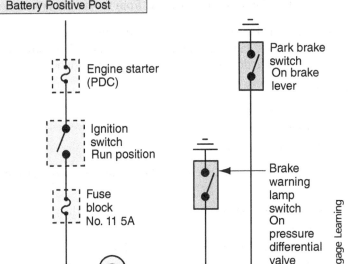

TASK E.5

29. The brake warning lamp pictured above stays on any time the ignition switch is on. Technician A says this could be caused by an open at the parking brake switch. Technician B says this could be caused by an open at the brake warning lamp switch. Who is correct?

    A. A
    B. B
    C. Both A and B
    D. Neither A nor B

    Answer A is incorrect. An open would prevent the light from coming on.

    Answer B is incorrect. An open would prevent the light from coming on.

    Answer C is incorrect. Both technicians are incorrect.

    Answer D is correct. A shorted circuit after the light could cause the light to stay on, not an open circuit.

30. Technician A says after bench bleeding the master cylinder it can be reinstalled. Technician B says after bench bleeding, the master cylinder should be emptied. Who is correct?

    A. A
    B. B
    C. Both A and B
    D. Neither A nor B

    TASK A.1.6

    Answer A is correct. Only Technician A is correct. After the master cylinder is bled it should be reinstalled on the vehicle.

    Answer B is incorrect. If the fluid was emptied out of the master cylinder it would defeat the purpose of bench bleeding the master cylinder.

    Answer C is incorrect. Only Technician A is correct.

    Answer D is incorrect. Technician A is correct.

31. While testing a vehicle equipped with a vacuum brake booster, if the engine is shut off and pressure applied to the brake pedal the pedal is hard and firm. Otherwise the brakes function normally. Which of the following is indicated by this test result?

    A. A restricted vacuum hose
    B. A leaking master cylinder
    C. A leaking vacuum check valve
    D. Air in the brake system

    TASK D.3

    Answer A is incorrect. A restricted vacuum hose would not allow vacuum to be applied to the booster, therefore the pedal would be hard all the time.

    Answer B is incorrect. A leaking master cylinder would cause a low, soft pedal.

    Answer C is correct. A leaking vacuum check valve would prevent vacuum being stored in the vacuum booster. There would be no reserve vacuum to act as a power assist for braking when the engine was shut off.

    Answer D is incorrect. Air in the brake system would cause a spongy pedal.

32. A vehicle had worn pads and they were replaced; now the vehicle's stopping power is poor. This is most likely caused by:

    A. Contaminated brake fluid.
    B. Air in the system.
    C. Improper break-in procedure.
    D. Faulty master cylinder.

    TASK C.15

    Answer A is incorrect. Replacing the pads would not have contaminated the brake fluid.

    Answer B is incorrect. Air in the system would cause a spongy pedal and is not a likely cause of poor stopping power after brake pad replacement.

    Answer C is correct. If the pads were not broken in properly after replacement, a lack of braking power could exist.

    Answer D is incorrect. A faulty master cylinder is not the most likely cause of poor braking after pad replacement.

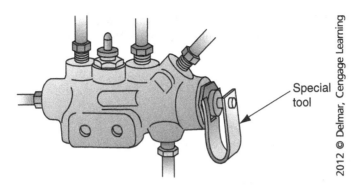

TASK A.4.2

33. The tool shown above is most likely used when:

    A. Bleeding the brakes.
    B. Bleeding the brake booster.
    C. Diagnosing a dash brake light concern.
    D. Diagnosing an ABS system concern.

    Answer A is correct. The tool is used to hold the combination valve open while bleeding the brakes.

    Answer B is incorrect. This tool is not used while working on a brake booster.

    Answer C is incorrect. This tool would not help to diagnose a brake warning light concern.

    Answer D is incorrect. This tool would not be helpful in diagnosing an ABS system concern.

TASK A.4.2

34. Which one of the following methods of brake bleeding typically requires two technicians?

    A. Gravity bleeding
    B. Brake pedal bleeding
    C. Vacuum bleeding
    D. Pressure bleeding

    Answer A is incorrect. Gravity bleeding is a one-technician operation.

    Answer B is correct. Bleeding the brakes using the brake pedal is usually a two-technician job.

    Answer C is incorrect. Vacuum bleeding is a one-technician job.

    Answer D is incorrect. Pressure bleeding using a bleeding ball is a one-technician job.

35. The hydro-boost belt driven pump is being tested. Which of the following would normally be considered the maximum pump pressure specification?

    A. 750 psi
    B. 1000 psi
    C. 1500 psi
    D. 2000 psi

    TASK D.4

    Answer A is incorrect. A reading of 750 psi could be caused by a slipping belt; 750 psi is much lower than the normal maximum pressure specification of 1500 psi.

    Answer B is incorrect. A reading of 1000 psi could be caused by a pressure relief valve opening prematurely; 1000 psi is lower than the normal maximum pressure specification of 1500 psi.

    Answer C is correct. A normal maximum pressure specification would be 1500 psi.

    Answer D is incorrect. A reading of 2000 psi would mean the pressure relief valve in the pump is not opening.

36. Which of the following would most commonly be used as the spindle nut on a set of tapered roller bearings?

    A. Self-locking nylon insert nut
    B. Castellated (slotted) nut with a cotter pin
    C. Double nuts with a lock between them
    D. Self-locking crimp style nut

    TASK E.1

    Answer A is incorrect. A nylon insert nut will sometimes be found on vehicle trim pieces, but not on the spindle of a set of tapered roller bearings.

    Answer B is correct. A castellated nut with a cotter pin is common so the wheel bearing clearance can be adjusted.

    Answer C is incorrect. Double nuts with a lock between them are sometimes used on the a full floating rear drive axle bearing set.

    Answer D is incorrect. A self-locking crimp style nut is not normally used. The castellated nut allows for the extra safety precaution of a cotter pin; a self locking crimp style nut does not.

Section 6 Answer Keys and Explanations — Brakes (A5)

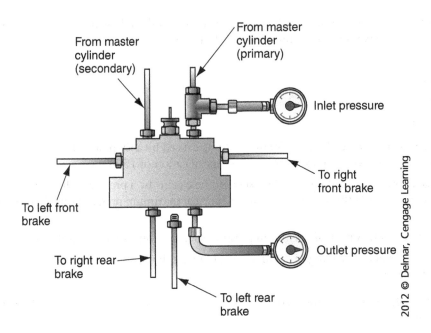

TASK A.3.2

37. After a brake application, the inlet pressure gauge in the figure above drops to normal; however, the outlet pressure gauge remains higher than normal. Technician A says a faulty master cylinder could be the cause. Technician B says a faulty parking brake could be the cause. Who is correct?

   A. A
   B. B
   C. Both A and B
   D. Neither A nor B

Answer A is incorrect. A master cylinder would not cause pressure to remain high on the inlet pressure gauge and not on the outlet pressure gauge. It is possible that a master which did not retract fully could cause both gauges to read high.

Answer B is incorrect. A faulty parking brake would not make the pressure remain high after a brake application.

Answer C is incorrect. Neither Technician is correct.

Answer D is correct. Neither Technician is correct. The most likely cause is a faulty combination valve.

TASK B.6

38. Brake shoe lubricant should be applied:

   A. To the shoe where it contacts the drum.
   B. To the shoe where it contacts the backing plate.
   C. To the drum where it contacts the hub.
   D. To the drum where it contacts the axle.

Answer A is incorrect. No lubricant is placed between the shoe and drum.

Answer B is correct. Lubricant should be installed between the shoe and backing plate.

Answer C is incorrect. No lubricant is placed between the drum and hub.

Answer D is incorrect. No lubricant is placed between the drum and axle.

Section 6 Answer Keys and Explanations

Brakes (A5)

39. A vehicle equipped with a hydro-boost system is leaking fluid where the booster and master cylinder meet. Which of the following is true?

    A. A leak here can only come from the booster assembly.
    B. A leak here can only come from the master cylinder.
    C. A pink fluid here would indicate a brake fluid leak.
    D. A pink fluid here would indicate a power steering fluid leak.

    TASK A.1.5

    Answer A is incorrect. The leak could be brake fluid from the master cylinder or power steering fluid from the booster.

    Answer B is incorrect. The fluid could be brake fluid from the master cylinder or power steering fluid from the booster.

    Answer C is incorrect. Brake fluid is clear or amber, not pink.

    Answer D is correct. A pink fluid would indicate power steering fluid, meaning the booster is leaking.

40. What measurement is performed on a disc brake rotor using a single dial indicator?

    A. Rotor lateral runout
    B. Rotor radial runout
    C. Rotor thickness variation
    D. Rotor imbalance

    TASK C.8

    Answer A is correct. Rotor lateral runout is measured with a single dial indiactor.

    Answer B is incorrect. Rotor radial runout is not a measurement.

    Answer C is incorrect. To measure rotor thickness variation with a dial indicator, two dial indicators must be used simutanously mounted directly across from each other.

    Answer D is incorrect. While it is possible that a rotor might be unbalanced and cause a concern, typically rotors do not become unbalanced and are not checked for balance. Most likely a technician would perform an on–the-car wheel balance to help compensate for an out of balance rotor.

41. While testing a brake system with a vacuum brake booster the technician finds that if the engine is started with pressure applied to the brake pedal the pedal will drop about 1/2 inch. Which of the following is indicated by this test result?

    A. A restricted vacuum hose
    B. A leaking master cylinder
    C. A bypassing master cylinder
    D. A normally operating booster

    TASK D.1

    Answer A is incorrect. A restricted vacuum hose would not allow vacuum to be applied to the booster; therefore, the pedal would not drop.

    Answer B is incorrect. A leaking master cylinder would cause a low, soft pedal.

    Answer C is incorrect. A bypassing master cylinder could cause the pedal to fade away while a constant pressure is applied.

    Answer D is correct. The brake booster is operating normally.

TASK A.4.1

42. Which of the following would be most likely to cause a brake fluid leak on a disc brake caliper assembly?

   A. Piston seal
   B. Piston boot
   C. Caliper slide pins
   D. Caliper slide bushings

   Answer A is correct. A faulty piston seal could cause a leak.

   Answer B is incorrect. A boot would not cause a leak; however, it can cause the piston to stick.

   Answer C is incorrect. The pins would not cause a leak; however, they can cause the brake to stick and drag.

   Answer D is incorrect. Bushings would not cause a leak; however, they can cause an uneven wear pattern on the brake pads.

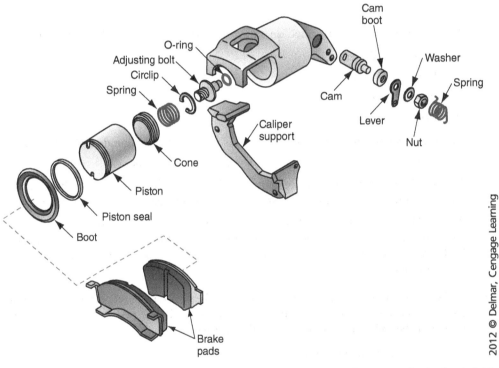

TASK E.3

43. Refer to the illustration above. The parking brake works correctly; however, the brake pedal is soft and spongy when the service brake pedal is pushed. Technician A says the problem could be a torn boot. Technician B says the problem could be air in the hydraulic system. Who is correct?

   A. A
   B. B
   C. Both A and B
   D. Neither A nor B

   Answer A is incorrect. A torn boot would affect both systems, because a torn boot can cause the piston to seize.

   Answer B is correct. Only Technician B is correct. Air in the hydraulic system would cause the service brakes to be spongy and soft.

   Answer C is incorrect. Only Technician B is correct

   Answer D is incorrect. Technician B is correct.

Section 6 Answer Keys and Explanations   **Brakes (A5)**

44. When installing a disc brake caliper boot the technician should stop driving when:
    A. The driving tool is flush with the caliper.
    B. The driving tool is below the caliper surface.
    C. The seal is flush with the caliper.
    D. The sound changes to a dull thud.

    TASK C.11

    Answer A is incorrect. Some boots will not install deeply enough to be flush. Attempting to drive them in until they are flush will damage them.

    Answer B is incorrect. Driving the boot in until the tool is below the caliper surface could force the boot too deeply into the caliper and damage the boot.

    Answer C is incorrect. The seal is not being installed.

    Answer D is correct. The technician should stop when the sound changes.

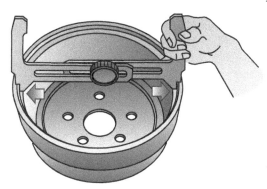

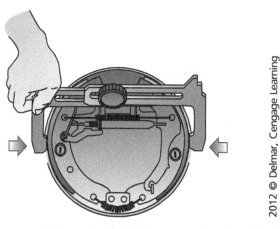

2012 © Delmar, Cengage Learning

45. Technician A says the above tool is used to install the springs on the shoes. Technician B says the tool shown above is used to measure the wear lip inside the drum. Who is correct?
    A. A
    B. B
    C. Both A and B
    D. Neither A nor B

    TASK B.9

    Answer A is incorrect. The tool is not used to install the springs on shoes.

    Answer B is incorrect. The tool is not used to measure the wear lip inside the drum.

    Answer C is incorrect. Neither Technician is correct.

    Answer D is correct. Neither Technician is correct. The tool is used to compare the outside diameter of the shoes to the inside diameter of the drum.

Section 6 Answer Keys and Explanations   Brakes (A5)

PREPARATION EXAM 5—ANSWER KEY

1. A	21. D	41. A
2. C	22. A	42. B
3. B	23. D	43. D
4. A	24. A	44. A
5. A	25. C	45. A
6. D	26. C	
7. B	27. D	
8. A	28. A	
9. A	29. A	
10. B	30. A	
11. B	31. A	
12. C	32. D	
13. B	33. A	
14. A	34. C	
15. B	35. D	
16. A	36. D	
17. B	37. D	
18. B	38. C	
19. A	39. A	
20. D	40. A	

PREPARATION EXAM 5—EXPLANATIONS

TASK C.8

1. Rotor lateral runout is greater than specification. This would most likely result in:
   A. Vibration while braking.
   B. Spongy brakes.
   C. A hard brake pedal.
   D. A low brake pedal.

   Answer A is correct. Excessive lateral runout will cause a vibration while braking.

   Answer B is incorrect. Spongy brakes are normally caused by air in the system.

   Answer C is incorrect. A hard brake pedal is usually caused by glazed linings or weak booster performance.

   Answer D is incorrect. A low brake pedal is normally caused by air in the system or misdajusted drum brakes.

Section 6 Answer Keys and Explanations

Brakes (A5)

2. Technician A says the parking brake adjustment may prevent the drum from fitting over new brake shoes. Technician B says some parking brakes are adjustable at the parking brake lever. Who is correct?

    A. A
    B. B
    C. Both A and B
    D. Neither A nor B

    TASK E.4

    Answer A is incorrect. Technician B is also correct.

    Answer B is incorrect. Technician A is also correct.

    Answer C is correct. Both Technicians are correct. If the parking brake adjustment has been tightened during brake service life and not returned to its original position during a brake shoe replacement, the new thicker linings can cause the drum to fail to fit over the new shoes. Additionally, after the initial brake adjustment at the drum brake, some parking brake cables are adjustable at the parking brake lever. This adjustment will compensate for a stretched cable.

    Answer D is incorrect. Both technicians are correct.

3. On a three-channel ABS system:

    A. The front wheels will share an ABS sensor.
    B. The rear wheels will share an ABS sensor.
    C. The left side will share an ABS sensor.
    D. The right side will share an ABS sensor.

    TASK F.4

    Answer A is incorrect. The front wheels never share a wheel speed sensor.

    Answer B is correct. On a three-channel ABS system the rear wheels may share a single wheel speed sensor. There are also some 4 sensor, 3 channel select low systems.

    Answer C is incorrect. Wheel speed sensors are not shared on the left side of a vehicle.

    Answer D is incorrect. Wheel speed sensors are not shared on the right side of a vehicle.

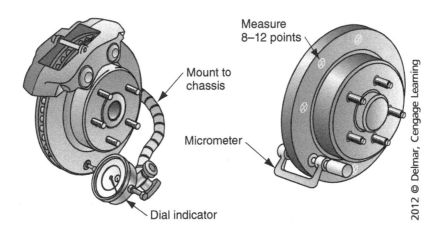

4. The micrometer above is measuring:

   A. Thickness variation.
   B. Lateral runout.
   C. Radial runout.
   D. Hardness.

   Answer A is correct. Thickness variation is being measured.

   Answer B is incorrect. Lateral runout is measured with a dial indicator.

   Answer C is incorrect. Radial runout is not usually measured on a disc brake rotor. However if it is, it is measured with a dial indiactor.

   Answer D is incorrect. Hardness of the disc brake rotor is not being measured. A Rockwell hardness test would be used to measure the hardness of the rotor; however, this test would not be practical to perform on a rotor.

5. The drum brake squeaks when the brakes are applied. Technician A says the shoes may be worn out. Technician B says the drum may be out of round. Who is correct?

   A. A
   B. B
   C. Both A and B
   D. Neither A nor B

   Answer A is correct. Only Technician A is correct. If the lining is worn to the shoe, the brakes will squeak.

   Answer B is incorrect. An out of round drum will cause a pedal pulsation of grabbing brake.

   Answer C is incorrect. Only Technician A is correct.

   Answer D is incorrect. Technician A is correct.

6. A car with a vacuum brake booster is idling too fast. When the vacuum line going to the brake booster is crimped shut the idle decreases to normal. Which of the following is the most likely cause?

   A. The vacuum check valve is leaking.
   B. The metering valve is leaking.
   C. The proportioning valve is leaking.
   D. The brake booster is leaking.

   TASK D.3

   Answer A is incorrect. A leaking vacuum check valve would cause no vacuum reserve when the engine is shut off.

   Answer B is incorrect. The metering valve is a hydraulic valve not a vacuum valve.

   Answer C is incorrect. The proportioning valve is a hydraulic valve not a vacuum valve.

   Answer D is correct. The brake booster is leaking vacuum and causing the high engine idle.

7. A vehicle pulls to the left while braking. All of the following could cause this EXCEPT:

   A. Sticking right front caliper piston.
   B. Sticking metering valve.
   C. Worn suspension components.
   D. Restricted right front brake line.

   TASK A.3.1

   Answer A is incorrect. A sticking right front caliper piston could cause the right front brake not to work, thus a pull to the left.

   Answer B is correct. A sticking metering valve would affect both front wheels.

   Answer C is incorrect. Worn suspension components can cause a shift in the alignment angles while braking; therefore, it would pull to the left while braking.

   Answer D is incorrect. A restricted right front brake line could cause the right front brake to be inoperative, thus a pull to the left.

8. The parking brake will not fully release. All of the following could be the cause EXCEPT:

   A. Air in the brake fluid.
   B. Binding parking brake cable.
   C. Broken return spring.
   D. Rusted actuator piston.

   TASK E.3

   Answer A is correct. Air in the brake fluid would not prevent the parking brake from releasing. Air in the brake fluid would allow the service brake to feel spongy.

   Answer B is incorrect. A binding cable can cause the parking brake to fail to release and drag.

   Answer C is incorrect. A broken return spring can cause the shoes to fail to return and the brake to drag.

   Answer D is incorrect. A rusted parking brake actuator can cause sticking and the brake to drag when the parking brake is released.

Section 6 Answer Keys and Explanations — Brakes (A5)

TASK C.8

9. The disc brake rotor thickness variation is greater than specification. This would most likely result in:

   A. Pedal pulsation.
   B. Spongy brakes.
   C. A hard brake pedal.
   D. A low brake pedal.

   Answer A is correct. Thickness variation would result in a pedal pulsation.

   Answer B is incorrect. Spongy brakes are normally caused by air in the system.

   Answer C is incorrect. A hard brake pedal is usually caused by glazed linings or weak booster performance.

   Answer D is incorrect. A low brake pedal is normally caused by air in the system or misadjusted drum brakes.

TASK A.3.3

10. The height sensing proportioning valve:

    A. Limits the rear brake pressure when the rear is fully loaded.
    B. Limits the rear brake pressure when the rear is lightly loaded.
    C. Limits the front brakes when the front is fully loaded.
    D. Limits the front brakes when the front is lightly loaded.

    Answer A is incorrect. The rear brakes need to be fully operational when the rear is fully loaded.

    Answer B is correct. During times of a light load in the rear the height sensing valve limits the pressure to the rear brakes to help prevent lockup.

    Answer C is incorrect. The front brakes are not controlled by the height sensing proportioning valve.

    Answer D is incorrect. The front brakes are not controlled by the height sensing proportioning valve.

TASK D.3

11. A vehicle is equipped with a vacuum brake booster. When the engine is running and the brakes are applied there is a constant sound of air movement (vacuum) around the brake pedal area. Technician A says the vacuum check valve is leaking. Technician B says the vacuum booster is leaking. Who is correct?

    A. A
    B. B
    C. Both A and B
    D. Neither A nor B

    Answer A is incorrect. A leaking vacuum check valve causes no vacuum reserve when the engine is shut off.

    Answer B is correct. Only Technician B is correct. The vacuum brake booster is defective and leaking vacuum.

    Answer C is incorrect. Only Technician B is incorrect.

    Answer D is incorrect. Technician B is correct.

12. Technician A says some wheel speed sensors are adjusted using a feeler gauge. Technician B says some ABS wheel speed sensors are adjusted using a paper spacer, which comes with the new wheel speed sensor. Who is correct?

    A. A
    B. B
    C. Both A and B
    D. Neither A nor B

    TASK F.7

    Answer A is incorrect. Technician B is also correct.

    Answer B is incorrect. Technician A is also correct.

    Answer C is correct. Both Technicians are correct. The wheel speed sensor may need to be adjusted by using a non-magnetic feeler gauge or a paper spacer.

    Answer D is incorrect. Both Technicians are correct.

13. The technician is replacing a section of steel brake line. The brake line should be cut with a:

    A. Hacksaw.
    B. Tubing cutter.
    C. Torch.
    D. Pipe cutter.

    TASK A.2.4

    Answer A is incorrect. A hacksaw would leave too rough of a cut.

    Answer B is correct. A tubing cutter is the correct tool.

    Answer C is incorrect. A torch would ruin the line.

    Answer D is incorrect. A pipe cutter would be too big.

14. Which of the following would be considered maximum thickness variation for a disc brake rotor?

    A. 0.0005″
    B. 0.002″
    C. 0.010″
    D. 0.025″

    TASK C.8

    Answer A is correct. This is an acceptable reading and generally considered to be the maximum acceptable reading.

    Answer B is incorrect. 0.002″ would be too much thickness variation on most vehicles.

    Answer C is incorrect. 0.010″ would be too much thickness variation.

    Answer D is incorrect. 0.025″ would be too much thickness variation.

Section 6 Answer Keys and Explanations — Brakes (A5)

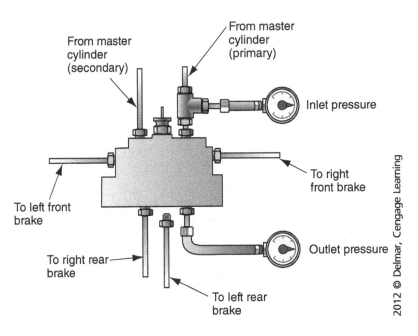

TASK A.4.1

15. The gauge pressures read normal during the initial test. After several high speed, high-pressure brake applications the gauges read lower and began to fluctuate. Which of the following could be the cause?

    A. The brake fluid is contaminated with oil.
    B. The brake fluid is contaminated with water.
    C. The master cylinder is faulty.
    D. The wheel cylinders are leaking.

    Answer A is incorrect. Brake fluid contaminated with oil will cause the rubber components to swell.

    Answer B is correct. The brake fluid is boiling. It is contaminated with water.

    Answer C is incorrect. A faulty master cylinder would not have good pressure in the first few tests.

    Answer D is incorrect. Leaking wheel cylinders will result in an external fluid leak and a low pedal.

TASK B.3

16. Which of the following would cause a glazed drum?

    A. A dragging shoe
    B. Air in the hydraulic system
    C. Missing brake shoe lining
    D. A frozen wheel cylinder

    Answer A is correct. A dragging shoe will cause the drum to overheat and glaze.

    Answer B is incorrect. Air in the system will cause a soft spongy pedal.

    Answer C is incorrect. A missing brake shoe lining will cause shoe to drum contact and a squeaking brake.

    Answer D is incorrect. A frozen wheel cylinder will cause an inoperative brake.

17. The parking brake will not apply. All of the following could be the cause EXCEPT:

    A. A broken parking brake cable.
    B. A broken return spring.
    C. A loose drum brake adjustment.
    D. A stretched parking brake cable.

    TASK E.4

    Answer A is incorrect. A broken parking brake cable will prevent the parking brake from applying.

    Answer B is correct. A broken return spring may cause brake drag, but would not cause failure of the parking brake to apply.

    Answer C is incorrect. A loose drum brake adjustment could result in too much clearance between the drum and shoes. This would result in the parking brake being unable to move the shoes sufficient distance to apply the brake.

    Answer D is incorrect. A stretched cable could prevent the shoes from travelling far enough to contact the drum.

18. Replacement metal brake lines should be constructed of:

    A. Copper.
    B. Steel.
    C. Aluminum.
    D. Brass.

    TASK A.2.4

    Answer A is incorrect. Copper is too soft.

    Answer B is correct. Metal brake lines should be constructed of steel.

    Answer C is incorrect. Aluminum is too soft.

    Answer D is incorrect. Brass is too soft.

19. Technician A says digital wheel speed sensors will have voltage supplied to them through the ABS control module. Technician B says digital wheel speed sensors are tested using an ohmmeter. Who is correct?

    A. A
    B. B
    C. Both A and B
    D. Neither A nor B

    TASK F.7

    Answer A is correct. Only Technician A is correct. Digital wheel speed sensors are supplied voltage from the ABS control module.

    Answer B is incorrect. Ohmmeter tests are performed on analog ABS wheel speed sensors.

    Answer C is incorrect. Only Technician A is correct.

    Answer D is incorrect. Technician A is correct.

Section 6 Answer Keys and Explanations — Brakes (A5)

TASK C.11

20. While installing the disc brake caliper boot it is ripped. Technician A says this will cause a brake fluid leak. Technician B says this will cause a low, soft brake pedal. Who is correct?

    A. A
    B. B
    C. Both A and B
    D. Neither A nor B

    Answer A is incorrect. A ripped boot will not cause a brake fluid leak. It does not seal brake fluid.

    Answer B is incorrect. A low, soft pedal is typically caused by air in the system. A ripped boot will not cause air in the brake fluid; the boot does not seal brake fluid.

    Answer C is incorrect. Neither technicain is correct.

    Answer D is correct. Neither technicain is correct. A ripped boot can cause the piston to stick after the vehicle is returned to service.

TASK A.1.6

21. A vehicle which has poor brake performance also has a high, hard brake pedal. Technician A says there may be air in the system. Technician B says the brakes may be out of adjustment. Who is correct?

    A. A
    B. B
    C. Both A and B
    D. Neither A nor B

    Answer A is incorrect. Air in the fluid causes a spongy pedal.

    Answer B is incorrect. Out of adjustment brakes will cause a low, firm brake pedal.

    Answer C is incorrect. Neither Technician is correct.

    Answer D is correct. Neither technician is correct. A high, hard, pedal can be caused by a non-functioning brake booster.

TASK F.8

22. Technician A says installing a tire with a different diameter than what was originally installed on the vehicle may set an ABS code. Technician B says using DOT 3 instead of DOT 4 brake fluid may set an ABS code. Who is correct?

    A. A
    B. B
    C. Both A and B
    D. Neither A nor B

    Answer A is correct. Only Technician A is correct. Tires of different diameters will have different rotational speeds; this will be recognized by the controller and can set an ABS code,

    Answer B is incorrect. DOT 3 and DOT 4 do have different specifications and may affect brake operation; however, it will not set an ABS code.

    Answer C is incorrect. Only Technician A is correct.

    Answer D is incorrect. Technician A is correct.

23. When the brake pedal is depressed, the rubber brake line connecting to the disc brake caliper swells. Which of the following is true?

    A. The caliper piston is stuck.
    B. The sliding caliper is stuck.
    C. The master cylinder is over pressurizing the hose.
    D. The hose should be replaced.

    TASK A.2.3

    Answer A is incorrect. A stuck caliper piston will prevent the brake from applying; however, the hose should be able to contain the hydraulic force without swelling.

    Answer B is incorrect. A stuck sliding caliper would prevent one disc brake pad from applying. This would result in poor brake performance, but the hose would still be able to contain the hydraulic pressure

    Answer C is incorrect. The hose is designed to withstand pressures greater than the master cylinder is able to produce.

    Answer D is correct. When the hose swells it is defective. A swollen hose will cause low pressure at the caliper and poor brake performance. It should be replaced.

24. Technician A says wheel bearings which are adjusted too tight can cause a wheel to lockup. Technician B says wheel bearings which are adjusted too tight will set an antilock braking system (ABS) trouble code. Who is correct?

    A. A
    B. B
    C. Both A and B
    D. Neither A nor B

    TASK E.1

    Answer A is correct. Only technician A is correct. Over tightening wheel bearings can cause them to lockup.

    Answer B is incorrect. Overtightened wheel bearings will not set an ABS trouble code.

    Answer C is incorrect. Only Technician A is correct.

    Answer D is incorrect. Technician A is correct.

TASK D.2

25. Which of the following could cause low vacuum to the booster with the engine running?
    A. A stuck open vacuum check valve
    B. A loose exhaust manifold
    C. A cracked intake manifold
    D. A stuck closed metering valve

    Answer A is incorrect. A stuck open vacuum check valve would cause no vacuum reserve when the engine was shut off.

    Answer B is incorrect. A loose exhaust manifold would not greatly affect the vacuum at the booster.

    Answer C is correct. A cracked intake manifold could cause low vacuum to the brake booster.

    Answer D is incorrect. A stuck closed metering valve could cause brake problems in the hydraulic circuit, but would not cause low vacuum to the booster.

TASK C.11

26. When installing a disc brake caliper boot which of the following is true concerning tool lubrication?
    A. The tool should be coated with brake fluid.
    B. The tool should be coated with engine oil.
    C. The tool should be dry.
    D. The tool should be coated with lubricating oil.

    Answer A is incorrect. The is no need for lubricant. Using brake fluid as a lubricant may cause a technician to diagnose a leak where one is not present.

    Answer B is incorrect. Using engine oil as a lubricant may result in the technician incorrectly identifying a leak.

    Answer C is correct. The tool should be dry.

    Answer D is incorrect. Using lubricating oil may result in the technician incorrectly identifying a leak.

Section 6 Answer Keys and Explanations

Brakes (A5)

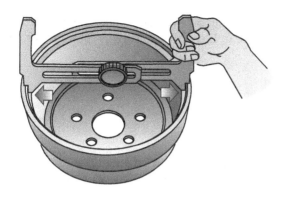

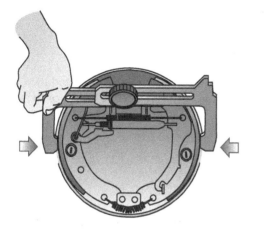

27. Technician A says the tool shown in the figure above is used to select the correct drum. Technician B says this tool is used to select the correct brake shoes. Who is correct?

   A. A
   B. B
   C. Both A and B
   D. Neither A nor B

TASK B.9

Answer A is incorrect. The technician determines which drum to use by part number. The technician should not use any drum other than the one recommended by the drum manufacturer.

Answer B is incorrect. The technician determines which brake shoe to use by part number. The technician should not use any brake shoes other than those recommended by the brake shoe manufacturer.

Answer C is incorrect. Neither Technician is correct.

Answer D is correct. Neither Technician is correct. The tool is used to compare the outside diameter of the shoes to the inside diameter of the drum.

Delmar, Cengage Learning ASE Test Preparation

TASK A.1.6

28. After several brake applications all the brakes drag on a vehicle. Technician A says the compensating port may be covered. Technician B says the replenishing port may be covered. Who is correct?

   A. A
   B. B
   C. Both A and B
   D. Neither A nor B

   Answer A is correct. Only Technician A is correct. If the compensating port is covered, the expanding fluid cannot move back-up into the master cylinder. Therefore the brakes will start to drag.

   Answer B is incorrect. The replenishing port will not cause the brakes to drag.

   Answer C is incorrect. Only Technician A is correct.

   Answer D is incorrect. Technician A is correct.

TASK F.9

29. A digital wheel speed sensor is being checked. When the sensor is disconnected the technician finds 5 V DC on the control module side of the connector. Which of the following is true?

   A. The voltage should be 12 V DC during this check.
   B. There should be no voltage present.
   C. There should be 2 V AC during this check.
   D. The test is being performed incorrectly.

   Answer A is correct. Digital wheel speed sensors will have 12 V DC supplied to them from the ABS control module.

   Answer B is incorrect. Digital wheel speed sensors need voltage supplied to them to generate a signal.

   Answer C is incorrect. The control module supplies 12 V DC to a digital wheel speed sensor.

   Answer D is incorrect. The test is being performed correctly.

Section 6 Answer Keys and Explanations

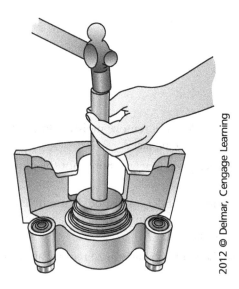

30. What is the technicain doing in the above illustration?

    A. Installing the boot
    B. Removing the piston
    C. Installing the piston
    D. Removing the boot

    TASK C.11

    Answer A is correct. The boot is being installed.

    Answer B is incorrect. The piston is not being removed. The piston is removed with air pressure.

    Answer C is incorrect. The piston is not driven in.

    Answer D is incorrect. The boot is removed with a seal removal tool.

31. The parking brake will not apply. Which of the following could be the cause?

    A. Misadjusted parking brake cable
    B. Misadjusted stop light switch
    C. Leaking vacuum booster check valve
    D. Leaking vacuum booster

    TASK E.3

    Answer A is correct. A misadjusted parking brake cable can result in insufficient travel of the brake shoes and failure of the shoes to engage the drum.

    Answer B is incorrect. A stop light switch can cause failure of the stop lights to illuminate or the stop lights to remain on; however, it would not affect the parking brake.

    Answer C is incorrect. The vacuum booster assists in the service brake application. A leaking vacuum booster does not affect parking brake operation.

    Answer D is incorrect. A leaking vacuum servo could prevent the parking brake from releasing, not applying.

TASK A.4.2

32. The purpose of bleeding the hydraulic system after a repair is:
    A. To remove moisture from the system.
    B. To flush old brake fluid from the system.
    C. To clean debris from the master cylinder.
    D. To remove air from the system.

Answer A is incorrect. Bleeding the master cylinder does not remove moisture from the hydraulic system. If moisture is present the fluid must be replaced.

Answer B is incorrect. Removing old brake fluid is a good practice, but is not the purpose of bleeding.

Answer C is incorrect. If there is debris in the master cylinder it must be disassembled, cleaned, and rebuilt. Normally in these circumstances the technician will replace the master cylinder; it is a more cost effective repair for the customer.

Answer D is correct. The purpose of bleeding the master cylinder is to remove air from the hydraulic system.

TASK C.8

33. The technician measures the rear rotors; the results are listed below:

    Minimum Rotor Thickness, Specification: 1.200″ (30.48mm)
    Actual Rotor Thickness, Left: 1.054″ (26.77mm)
    Actual Rotor Thickness, Right: 1.230″ (31.24mm)

    Which of the following should be done?
    A. Replace the left rotor.
    B. Replace both rotors.
    C. Reuse both rotors.
    D. Replace the right rotor.

Answer A is correct. The left rotor is out of specification.

Answer B is incorrect. The right rotor is within specification.

Answer C is incorrect. The left rotor is not reusable.

Answer D is incorrect. The right rotor is within specification.

TASK A.4.2

34. When bleeding the brakes using the brake pedal, which of the following is true?
    A. Loosen the bleeder screw and pump the pedal.
    B. Make quick pedal applications then loosen the bleeder screw.
    C. Slowly apply the brake pedal, closing the bleeder screw before the end of the stroke.
    D. Stroke the pedal all the way to the floor.

Answer A is incorrect. This would pull air back into the system.

Answer B is incorrect. This will cause big bubbles to break into small bubbles and be harder to remove.

Answer C is correct. Slow smooth strokes work best to bleed the brakes.

Answer D is incorrect. If the pedal is stroked all the way to the floor, the cups in the master cylinder may be cut by corrosion in the end of the bore.

Section 6 Answer Keys and Explanations

Brakes (A5)

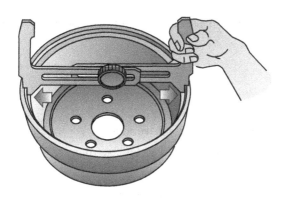

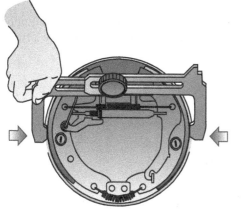

35. After the tool shown in the figure above is sized to the drum, it will not fit over the shoes. What should the technician do?

    A. Replace the shoes.
    B. Turn the drum.
    C. Replace the springs.
    D. Adjust the shoes.

    TASK B.9

    Answer A is incorrect. The shoes do not need to be replaced, they need to be adjusted.

    Answer B is incorrect. The drum does not need to be turned; the shoes need to be adjusted.

    Answer C is incorrect. This tool does not determine the condition of the springs.

    Answer D is correct. The shoes need to be adjusted.

36. Technician A says the parking brakes should be adjusted before the service brakes. Technician B says that a stretched parking brake cable can cause the parking brake to drag. Who is correct?

    A. A
    B. B
    C. Both A and B
    D. Neither A nor B

    TASK E.4

    Answer A is incorrect. The service brake adjustment should be performed prior to the parking brake adjustment. This ensures the base foundation brake is functioning properly.

    Answer B is incorrect. A stretched parking brake cable can cause the brake to fail to apply, not drag.

    Answer C is incorrect. Neither Technician is correct.

    Answer D is correct. Neither Technician is correct.

TASK C.8

37. Which of the following would indicate maximum allowable runout of a disc brake rotor?

    A. 0.0005″
    B. 0.0002″
    C. 0.0004″
    D. 0.0020″

    Answer A is incorrect. This would be considered acceptable runout.

    Answer B is incorrect. This would generally be considered acceptable runout.

    Answer C is incorrect. This would be acceptable runout.

    Answer D is correct. This is the typically considered the maximum rotor runout specification.

TASK F.5

38. Technician A says a scan tool may be necessary to bleed the brakes on an ABS equipped vehicle. Technician B says bleeding brakes on an ABS equipped vehicle may be a two-technician job. Who is correct?

    A. A
    B. B
    C. Both A and B
    D. Neither A nor B

    Answer A is incorrect. Technician B is also correct.

    Answer B is incorrect. Technician A is also correct.

    Answer C is correct. Both Technicians are correct. When bleeding ABS brakes, most manufacturers recommend pulsing the solenoid valves during the bleeding procedure. The scan tool will be needed to pulse the valves. If a pressure or vacuum bleeder is not used, then two technicians will most likely be needed to bleed the brakes.

    Answer D is incorrect. Both Technicians are correct.

TASK A.1.5

39. When the brake pedal is depressed fluid leaks from the rear of the master cylinder. This could be caused by:

    A. A leaking primary piston seal.
    B. A leaking secondary piston seal.
    C. A leaking primary cup.
    D. A leaking secondary cup.

    Answer A is correct. The primary piston seal is leaking.

    Answer B is incorrect. If the secondary seal leaks the fluid will pass internally and not leak externally.

    Answer C is incorrect. A leaking primary cup bypasses into the reservoir.

    Answer D is incorrect. A leaking secondary cup bypasses into the reservoir. This is an internal leak which may cause a low pedal, but would not cause an external leak.

Section 6 Answer Keys and Explanations   Brakes (A5)

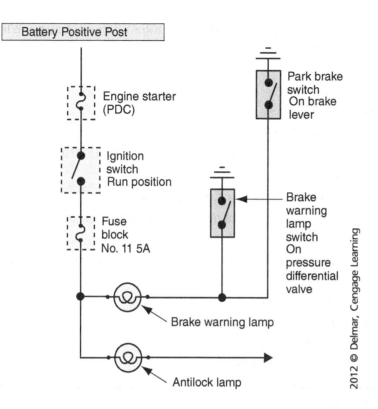

40. The brake warning lamp pictured above will occasionally come on then later go off. Which of the following could be the cause?

    A. A misadjusted park brake switch
    B. Water contaminated brake fluid
    C. Worn parking brakes
    D. Worn brake drums

    TASK E.5

    Answer A is correct. A misadjusted switch could allow the parking brake switch to move away from the lever occasionally and the light to come on.

    Answer B is incorrect. The master cylinder switch does not detect water in the brake fluid.

    Answer C is incorrect. Worn parking brakes would not cause the light to come on.

    Answer D is incorrect. Worn brake drums would not cause the light to come on.

41. The brake pedal has a pulsation when the brakes are applied. Technician A says the drum may be out of round. Technician B says the drum may be glazed. Who is correct?

    A. A
    B. B
    C. Both A and B
    D. Neither A nor B

    TASK B.2

    Answer A is correct. Only Technician A is correct. An out of round drum may cause a pulsation.

    Answer B is incorrect. A glazed drum will cause a hard pedal with poor brake performance.

    Answer C is incorrect. Only Technician A is correct.

    Answer D is incorrect. Technician A is correct.

TASK C.8

42. The technician makes one measurement on a disc brake rotor with a micrometer. Which of the following is being measured?

   A. Brake rotor diameter
   B. Brake rotor thickness
   C. Brake rotor lateral runout
   D. Brake rotor thickness variation

   Answer A is incorrect. Brake rotor diameter is typically measured with a tape measure.

   Answer B is correct. Brake rotor thickness is being measured.

   Answer C is incorrect. To measure runout a dial indicator is used.

   Answer D is incorrect. Rotor thickness variation requires several measurements to be taken around the rotor.

TASK F.4

43. The ABS controller has stored a code for the left front wheel speed sensor. All of the following could be the cause EXCEPT:

   A. A wheel speed sensor resistance reading of 0.2 ohms.
   B. Mismatched tire diameters.
   C. A loose wheel bearing on the left front.
   D. A loose wheel bearing on the right front.

   Answer A is incorrect. A resistance reading of 0.2 ohms would indicate a shorted wheel speed sensor. This would set a code.

   Answer B is incorrect. Mismatched tire diameters would indicate different wheel speeds on the vehicle. The controller could interpret this as a faulty wheel speed sensor and set a diagnostic trouble code.

   Answer C is incorrect. A loose wheel bearing can cause the clearance between the tone ring and sensor to change, resulting in a faulty wheel speed sensor signal. This would set a code for the left front wheel speed sensor.

   Answer D is correct. The code is for the left front not the right front. A loose wheel bearing on the right front would cause an ABS wheel speed sensor code for the right front, not the left front.

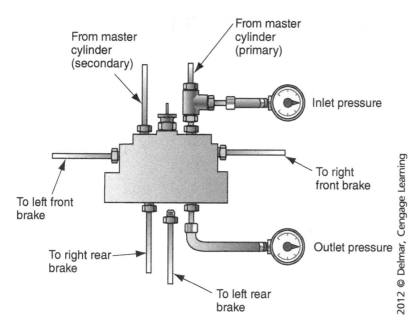

44. During a pressure test both gauges read lower than normal. Technician A says low fluid level in the primary reservoir could be the cause. Technician B says low fluid level in the secondary reservoir could be the cause. Who is correct?

   A. A
   B. B
   C. Both A and B
   D. Neither A nor B

   TASK A.3.1

   Answer A is correct. Only Technician A is correct. The gauges are connected to the primary circuit; low fluid level in the primary reservoir could cause these pressures to be lower than normal.

   Answer B is incorrect. The gauges are connected to the primary side. Secondary fluid level would not affect the reading.

   Answer C is incorrect. Only Technician A is correct.

   Answer D is incorrect. Technician A is correct.

45. While testing the vacuum brake booster the technician finds that if the engine is started with pressure applied to the brake pedal the pedal does not move. Which of the following is indicated by this test result?

   A. A restricted vacuum hose
   B. A leaking master cylinder
   C. A bypassing master cylinder
   D. A normally operating booster

   TASK D.1

   Answer A is correct. A restricted vacuum hose would not allow vacuum to be applied to the booster; therefore, the pedal would not drop.

   Answer B is incorrect. A leaking master cylinder would cause a low, soft pedal.

   Answer C is incorrect. A bypassing master cylinder could cause the pedal to fade away while a constant pressure is applied.

   Answer D is incorrect. A normally operating brake booster will cause the pedal to fall slightly when the engine is started.

Section 6 Answer Keys and Explanations   Brakes (A5)

PREPARATION EXAM 6—ANSWER KEY

1. C	21. C	41. C
2. D	22. A	42. C
3. B	23. A	43. D
4. A	24. A	44. D
5. D	25. A	45. C
6. C	26. C	
7. B	27. C	
8. A	28. C	
9. C	29. C	
10. A	30. D	
11. B	31. C	
12. D	32. B	
13. C	33. C	
14. A	34. C	
15. A	35. B	
16. A	36. C	
17. D	37. D	
18. A	38. D	
19. A	39. C	
20. D	40. B	

PREPARATION EXAM 6—EXPLANATIONS

TASK C.3

1. What would a C-clamp be used for during disc brake service?

   A. Seating the pads on the rotor
   B. Seating the caliper on the guides
   C. Retracting the piston in the bore
   D. Compressing the pads into the piston

   Answer A is incorrect. The pads are seated to the rotor during brake burnishing. A C-clamp is not needed for this procedure.

   Answer B is incorrect. A tool is not necessary to seat the caliper on the guides. It is done by hand.

   Answer C is correct. A C-clamp is used to retract the disc brake caliper piston into the bore.

   Answer D is incorrect. The pads do not need to be compressed into the piston.

2. The red brake warning light illuminates when the brake pedal is depressed. Which of the following is the LEAST LIKELY cause?

   A. Air in the front brake system
   B. Air in the rear brake system
   C. Low brake fluid level
   D. Faulty ABS pump assembly

   TASKS A.3.4, F.4

   Answer A is incorrect. Air in the front brake system can cause the pressure differential switch to light the red brake warning light.

   Answer B is incorrect. Air in the rear brake system can cause the pressure differential switch to light the red brake warning light.

   Answer C is incorrect. Low brake fluid can cause the red brake warning light to illuminate.

   Answer D is correct. A faulty ABS pump assembly would light the ABS warning light.

3. Which of the following would be considered the most normal brake fluid service interval?

   A. 1 year/15,000 miles
   B. 3 year/36,000 miles
   C. 6 years/72,000 miles
   D. 10 years/100,000 miles

   TASK A.4.2

   Answer A is incorrect. This is not a normal service interval. This is more often than most manufacturers recommend.

   Answer B is correct. Although a few manufacturers do not specify a service interval, of those that do the most common recommendation is 3 years/36,000 miles.

   Answer C is incorrect. This is longer than normal.

   Answer D is incorrect. This is longer than normal.

4. The dimension stamp on a rotor is 1.18" (30.00mm). The left front rotor measures 1.17" (31.75mm). The right front rotor measures 1.29" (29.72). Which of the following is true?

   A. The left rotor needs to be replaced.
   B. The right rotor needs to be replaced.
   C. Both rotors need to be replaced.
   D. Neither rotor needs to be replaced.

   TASK C.8

   Answer A is correct. The left rotor is below minimum specifications and needs to be replaced.

   Answer B is incorrect. The right rotor is thicker than specifications and does not need to be replaced.

   Answer C is incorrect. The right rotor is within specifications.

   Answer D is incorrect. The left rotor needs to be replaced.

TASK A.4.4

5. The brake fluid in a vehicle is black. Which of the following is the most likely cause?

   A. The brake fluid is contaminated with power steering fluid.
   B. The brake fluid is contaminated with water.
   C. Black is a normal brake fluid color.
   D. The brake fluid needs to be replaced with new.

   Answer A is incorrect. Brake fluid contaminated with power steering fluid is not black. It will cause swollen rubber swells in the master cylinder reservoir cap.

   Answer B is incorrect. Water contaminated brake fluid is not black. Water contaminated brake fluid can be identified using test strips.

   Answer C is incorrect. Brake fluid is not black. Normally brake fluid is clear to amber.

   Answer D is correct. The brake fluid needs to be replaced with new. Black brake fluid indicates old brake fluid.

TASK E.4

6. Technican A says a 1/4" drill bit may be needed to check the parking brake adjustment on some vehicles. Technician B says a hose clamp can be used to release the parking brake cable from the backing plate. Who is correct?

   A. A
   B. B
   C. Both A and B
   D. Neither A nor B

   Answer A is incorrect. Technician B is also correct.

   Answer B is incorrect. Technician A is also correct.

   Answer C is correct. Both Technicians are correct. A drill bit is involved in some parking brake adjustment sequences, and a hose clamp can help depress the locking tangs on a parking brake cable.

   Answer D is incorrect. Both Technicians are correct.

TASK D.4

7. The brake pedal vibrates when applied on a vehicle with a hydro-boost brake booster. Which of the following is the most likely cause?

   A. A restricted booster supply hose
   B. A loose power steering pump belt
   C. A restricted front brake line
   D. A restricted rear brake line

   Answer A is incorrect. A restricted booster supply line could cause poor booster performance but not a vibration.

   Answer B is correct. A loose power steering pump belt could slip when the brakes are applied and the pump is required to make pressure. This would result in a vibration in the brake pedal.

   Answer C is incorrect. A restricted front brake line would cause the front brake to be slow to apply and release but would not cause a pedal vibration.

   Answer D is incorrect. A restricted rear brake line would cause the rear brake to be slow to apply and release but would not cause the brake pedal to vibrate.

8. When bleeding the brakes using the brake pedal, which of the following is true?

   A. A hose should be attached to the bleeder screw and submerged in brake fluid.
   B. The bleeder screw should be closed throughout the downward stroke of the brake pedal.
   C. The bleeder screw should be loosened two complete turns.
   D. The brake pedal should be stroked as quickly as possible.

   TASK A.4.2

   Answer A is correct. This method will help prevent drawing air back into the system.

   Answer B is incorrect. The bleeder screw should be open during the down stroke.

   Answer C is incorrect. A quarter turn is all that is necessary.

   Answer D is incorrect. The brake pedal should be stroked slowly.

9. Technician A says the traction control system can signal the engine control module to reduce engine torque. Technician B says the traction control system can apply the vehicle brakes. Who is correct?

   A. A
   B. B
   C. Both A and B
   D. Neither A nor B

   TASK F.1

   Answer A is incorrect. Technician B is also correct.

   Answer B is incorrect. Technician A is also correct.

   Answer C is correct. Both Technicians are correct. Both methods are used by the traction control system to regain traction during a wheel spin.

   Answer D is incorrect. Both Technicians are correct.

10. The specification stamped on the drum is 8.125" (206.38mm). Which measurement below would indicate a re-useable drum?

    A. 8.025" (203.84mm)
    B. 8 .126" (206.38mm)
    C. 8.225" (208.92mm)
    D. D.8.325" (211.45mm)

    TASK B.3

    Answer A is correct. This drum measurement is under the maximum dimension.

    Answer B is incorrect. This measurement is above the maximum dimension. It would be illegal to install a drum on a vehicle that is over the specification stamped on the drum. .

    Answer C is incorrect. This drum is oversized.

    Answer D is incorrect. This drum is oversized.

TASK A.3.1

11. The rear brakes tend to lock during hard braking. Which of the following could be the cause?

    A. Metering valve
    B. Proportioning valve
    C. Quick take up valve
    D. Pressure differential valve

    Answer A is incorrect. A failed metering valve may cause the front wheels to lock on slippery surfaces; but would not affect the rear brakes.

    Answer B is correct. A malfunctioning proportioning valve could cause the rear brakes to get too much pressure in a hard brake application and the rear brakes to lock.

    Answer C is incorrect. The quick take up valve will not cause this situation. It could cause a low brake pedal.

    Answer D is incorrect. The pressure differential valve will not cause this condition. It could cause the brake warning light to stay on.

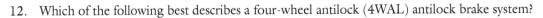
TASK F.1

12. Which of the following best describes a four-wheel antilock (4WAL) antilock brake system?

    A. It is a four-wheel antilock system used only on four-wheel drive vehicles.
    B. It is a four-wheel antilock system used only on all-wheel drive vehicles.
    C. It is a four-wheel antilock system used only on pick-up trucks.
    D. It is a four-wheel antilock system used on a variety of vehicles.

    Answer A is incorrect. 4WAL is used on many vehicles, not only four-wheel drive vehicles.

    Answer B is incorrect. 4WAL is not used solely on all-wheel drive vehicles.

    Answer C is incorrect. Rear-wheel antilock (RWAL) is typically used on pick-up trucks.

    Answer D is correct. 4WAL is a four-wheel antilock system used on a variety of vehicles.

TASK C.9

13. Rotor runout is checked with a dial indicator and found to be out of specification. The rotor is removed, rotated 180 degrees, and reinstalled. The runout is checked again and is still out of specification. The excessive runout is in the same location on the hub. Which of the following is indicated?

    A. The measurement is being performed incorrectly.
    B. The dial indicator is faulty.
    C. The runout is in the hub.
    D. The runout is in the rotor.

    Answer A is incorrect. The measurement is being performed correctly to determine if the runout is in the hub or rotor.

    Answer B is incorrect. There is no reason to suspect the dial indicator is faulty.

    Answer C is correct. The runout did not move with the rotor; therefore, it must be in the hub.

    Answer D is incorrect. If the runout was in the rotor the location with excessive runout would have moved with the rotor and not with the hub.

14. Which of the following would cause a drum to have hard spots?

    A. A broken return spring
    B. Air in the hydraulic system
    C. Missing brake shoe lining
    D. A frozen wheel cylinder

    TASK B.2

    Answer A is correct. A broken return spring will cause the brake to drag and the drum to overheat, resulting in drum hard spots.

    Answer B is incorrect. Air in the system will cause a spongy pedal.

    Answer C is incorrect. A missing lining would cause a grooved drum.

    Answer D is incorrect. A frozen wheel cylinder will cause an inoperative brake.

15. When the brakes are applied at low speed on a slippery surface the front brakes lock. Which of the following could be the cause?

    A. Metering valve
    B. Proportioning valve
    C. Quick take up valve
    D. Pressure differential valve

    TASK A.3.1

    Answer A is correct. A failed metering valve may let the front brakes apply to soon. This could cause the front wheels to lock on slippery surfaces.

    Answer B is incorrect. A malfunctioning proportioning valve could cause the rear brakes to lock, but would not affect the front brakes.

    Answer C is incorrect. A failed quick take up valve would not lock the front brakes. It could cause a low brake pedal.

    Answer D is incorrect. A failed pressure differential valve would not lock the front brakes. It could cause the brake warning light to stay on.

16. The parking brake warning light switch is misadjusted. Which light on the dash will most likely illuminate?

    A. Red brake warning light
    B. Amber brake warning light
    C. ABS light
    D. Traction control (TRAC) light

    TASK E.5

    Answer A is correct. The parking brake activates the red brake warning light.

    Answer B is incorrect. The amber light is usually the ABS light, not a brake light.

    Answer C is incorrect. The ABS light is usually amber, and would not be activated by a problem with the brake warning light switch. The parking brake warning light is red.

    Answer D is incorrect. The traction control light would not be illuminated.

TASK A.4.4

17. Which of the following is the correct disposal method for used brake fluid?

   A. Pour it down the drain.
   B. Pour it in with used oil to be recycled.
   C. Pour it on the ground.
   D. Save it in a separate container for recycling.

   Answer A is incorrect. Used brake fluid should not be poured down the drain.

   Answer B is incorrect. Used brake fluid should not be mixed with used engine oil for recycling.

   Answer C is incorrect. Used brake fluid should not be poured on the ground.

   Answer D is correct. Used brake fluid should be stored in its own container for recycling.

TASK C.10

18. Technician A says machining the rotor on the vehicle helps to compensate for hub runout. Technician B says machining the rotor on the vehicle helps to compensate for wheel runout. Who is correct?

   A. A
   B. B
   C. Both A and B
   D. Neither A nor B

   Answer A is correct. Only Technician A is correct. One advantage of machining a rotor on the car that any rotor runout in the hub is compensated for.

   Answer B is incorrect. Wheel runout will not be compensated for by turning the rotors on the car. However hub runout is compensated for.

   Answer C is incorrect. Only Technician A is correct.

   Answer D is incorrect. Technician A is correct.

TASK F.1

19. Which of the following best describes a three channel ABS system?

   A. The rear wheels share a circuit.
   B. The front wheels share a circuit.
   C. The left side shares a circuit.
   D. The right side shares a circuit.

   Answer A is correct. The front wheels operate individually of each other in a three channel ABS system, and the rear wheels operate as a single unit.

   Answer B is incorrect. No ABS system has the front wheels share a circuit.

   Answer C is incorrect. No automotive ABS systems have the left side share a circuit.

   Answer D is incorrect. No automotive ABS systems have the right side share a circuit.

Section 6 Answer Keys and Explanations

Brakes (A5)

20. Technician A says DOT 5 brake fluid is clear. Technician B says DOT 3 brake fluid is purple. Who is correct?

    A. A
    B. B
    C. Both A and B
    D. Neither A nor B

    Answer A is incorrect. DOT 5 brake fluid is purple.

    Answer B is incorrect. DOT 3 brake fluid is clear or amber.

    Answer C is incorrect. Neither Technician is correct.

    Answer D is correct. Neither Technician is correct.

    TASK A.2.3

21. Technician A says the red brake warning light is turned on by the parking brake warning light switch. Technician B says the red brake warning light is turned on by the pressure differential switch. Who is correct?

    A. A
    B. B
    C. Both A and B
    D. Neither A nor B

    Answer A is incorrect. Technician B is also correct.

    Answer B is incorrect. Technician A is also correct.

    Answer C is correct. Both Technicians are correct. The red warning lamp can be activated by the pressure differential switch or the parking brake switch.

    Answer D is incorrect. Both Technicians are correct.

    TASK E.5

22. What is meant by "indexing" a rotor?

    A. Matching the high spot on the rotor and low spot on the hub
    B. Matching the low spot on the rotor and the low spot on the hub
    C. Matching the high spot on the rotor and the high spot on the hub
    D. Matching the widest spot on the rotor to the widest spot on the hub

    Answer A is correct. Indexing a rotor means putting the high and low spot together to create the lowest possible runout.

    Answer B is incorrect. Putting the low and low together would result in increasing total runout.

    Answer C is incorrect. Indexing a rotor means putting the high and low together to lower the total runout. Putting the high spot on the rotor and the high spot on the hub together would result in increasing runout.

    Answer D is incorrect. Indexing a rotor is an attempt to lower total runout to an acceptable amount. The width is not a factor.

    TASK C.8

Section 6 Answer Keys and Explanations

Brakes (A5)

TASK C.1

23. On a vehicle with disc brakes, the outside pad is worn much more than the inside pad. Which of the following could be the cause?

   A. Sticking caliper slides
   B. Collapsed brake line
   C. A sticking metering valve
   D. Swollen brake line

   Answer A is correct. Sticking caliper slides can cause the brake to fail to fully release; this could cause the outer pad to drag and wear quickly.

   Answer B is incorrect. A collapsed brake line can cause the brake to drag, but both pads would drag equally.

   Answer C is incorrect. A sticking metering valve may cause the brakes to grab, but would not cause the outside pad to wear more than the inside pad.

   Answer D is incorrect. A swollen brake line can cause a low brake pedal but would not cause uneven pad wear.

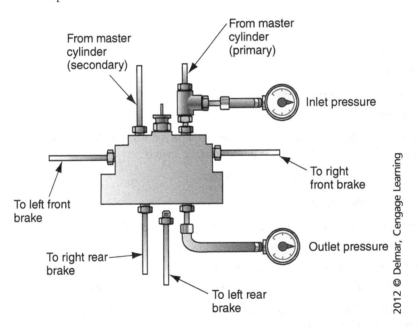

TASKS A.3.2, F.4

24. After a brake application, the inlet pressure gauge drops to normal; however, the outlet pressure gauge remains higher than normal. Technician A says a faulty combination valve could be the cause. Technician B says a faulty antilock braking system (ABS) modulator valve could be the cause. Who is correct?

   A. A
   B. B
   C. Both A and B
   D. Neither A nor B

   Answer A is correct. A restricted combination valve could hold pressure at the outlet pressure gauge.

   Answer B is incorrect. A faulty ABS modulator valve would not make the pressure at the gauge remain high after a brake application. If it were leaking it could cause the pressure to be low.

   Answer C is incorrect. Only Technician A is correct.

   Answer D is incorrect. Technician A is correct.

Page 218

Delmar, Cengage Learning ASE Test Preparation

Section 6 Answer Keys and Explanations — Brakes (A5)

25. A parking brake drags after application and release. Technician A says the cable may be rusty. Technician B says the brake drum may be bell mouthed. Who is correct?

    A. A
    B. B
    C. Both A and B
    D. Neither A nor B

    Answer A is correct. Only Technician A is correct. A rusty parking brake cable could stick and cause the brakes to drag.

    Answer B is incorrect. A bell mouthed brake drum would not cause a parking brake to fail to release.

    Answer C is incorrect. Only Technician A is correct.

    Answer D is incorrect. Technician A is correct.

26. Technician A says some vacuum brake booster systems used on diesel engines will have an engine driven vacuum pump. Technician B says some vacuum brake booster systems used on gasoline engines will have an engine driven vacuum pump. Who is correct?

    A. A
    B. B
    C. Both A and B
    D. Neither A nor B

    Answer A is incorrect. Technician B is also correct.

    Answer B is incorrect. Technician A is also correct.

    Answer C is correct. Engine driven vacuum pumps are used on diesel powered vehicles, as well as some gasoline powered vehicles, which have low manifold vacuum.

    Answer D is incorrect. Both technicians are correct.

27. Which of the following tools would most likely be used to measure wheel bearing end-play?

    A. Outside micrometer
    B. Inside micrometer
    C. Dial indicator
    D. Feeler gauge

    Answer A is incorrect. An outside micrometer can be used to measure rotor thickness; wheel bearing end-play cannot be measured with a micrometer.

    Answer B is incorrect. An inside micrometer can be used to measure drum diameter; wheel bearing end-play cannot be measured with a micrometer.

    Answer C is correct. A dial indicator is the correct tool to measure the end-play of a wheel bearing.

    Answer D is incorrect. A feeler gauge is not an effective method to measure wheel bearing end-play. A feeler gauge can be used to measure spark plug gap.

28. Which of the following would cause a grooved drum?

    A. A dragging shoe
    B. Air in the hydraulic system
    C. Missing brake shoe lining
    D. A frozen parking brake cable

    Answer A is incorrect. A dragging shoe will cause a glazed drum.

    Answer B is incorrect. Air will cause a spongy pedal.

    Answer C is correct. A missing lining can cause the brake shoe to groove the drum.

    Answer D is incorrect. A frozen cable would affect the parking brake. It could cause the brake to fail to apply or to stick after it was applied. If it sticks and the vehicle is driven, the brake shoes will become glazed from overheating.

29. Technicain A says excessive rotor runout can be caused by a warped disc brake rotor. Technician B says excessive rotor runout can be caused by hub runout. Who is correct?

    A. A
    B. B
    C. Both A and B
    D. Neither A nor B

    Answer A is incorrect. Technician B is also correct.

    Answer B is incorrect. Technician A is also correct.

    Answer C is correct. Both Technicians are correct. Excessive rotor runout can be caused by a warped disc brake rotor or excessive runout in the hub.

    Answer D is incorrect. Both Technicians are correct.

30. When constant pressure is held on the brake pedal it will slowly fade to the floor. Which is the most likely cause?

    A. Air in the brake fluid
    B. Stuck caliper piston
    C. Stuck wheel cylinder
    D. Internally leaking master cylinder

    Answer A is incorrect. Air in the brake fluid would cause a spongy pedal but would not cause the pedal to fade to the floor.

    Answer B is incorrect. A stuck caliper piston would cause a hard pedal.

    Answer C is incorrect. A stuck wheel cylinder would cause a hard pedal.

    Answer D is correct. A master cylinder that is leaking internally would allow fluid to bypass, causing the pedal to fall slowly to the floor.

Section 6 Answer Keys and Explanations

Brakes (A5)

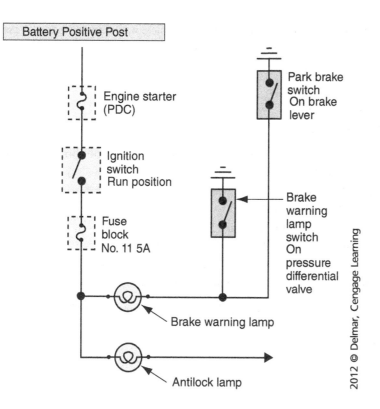

31. The brake warning lamp pictured above stays on any time the ignition switch is on. Technician A says this could be caused by a stuck closed parking brake switch. Technician B says this could be caused by a stuck closed brake warning lamp switch. Who is correct?

    A. A
    B. B
    C. Both A and B
    D. Neither A nor B

TASK E.5

Answer A is incorrect. Technician B is also correct.

Answer B is incorrect. Technician A is also correct.

Answer C is correct. Both technicians are correct. If either switch was stuck closed the light would remain on any time the ignition switch was on.

Answer D is incorrect. Both Technicians are correct.

Section 6 Answer Keys and Explanations   Brakes (A5)

TASK D.4

32. The brake pedal is slow to return when released on a vehicle with a hydro-boost brake booster. Which of the following is the most likely cause?

    A. A restricted booster supply hose
    B. A restricted booster return hose
    C. A restricted front brake line
    D. A restricted rear brake line

    Answer A is incorrect. A restricted booster supply line could cause poor booster performance but not cause a slow pedal return.

    Answer B is correct. A restricted booster return line can cause the booster to be slow to release and a slow return on the brake pedal.

    Answer C is incorrect. A restricted front brake line would cause the front brake to be slow to apply and release but would not affect the brake pedal return.

    Answer D is incorrect. A restricted rear brake line would cause the rear brake to be slow to apply and release but would not affect the brake pedal return.

TASK C.9

33. Technician A says some disc brake rotors are directional and can be installed only on one side of a car. Technician B says rotor thickness variation and parallelism refer to the same measurement. Who is correct?

    A. A
    B. B
    C. Both A and B
    D. Neither A nor B

    Answer A is incorrect. Technician B is also correct.

    Answer B is incorrect. Technician A is also correct.

    Answer C is correct. Both Technicians are correct. Some rotors have directional vents; installing on the wrong side will cause rotor overheating. Thickness variation and parallelism both refer to difference in the thickness of the rotor when measured in several locations around the rotor.

    Answer D is incorrect. Both Technicians are correct.

TASK F.7

34. The wheels speed sensor is being checked with an ohmmeter. When the technician measures between one of the wires and chassis ground the ohmmeter shows 0.2 ohms. This indicates:

    A. The sensor is open.
    B. The sensor is shorted to voltage.
    C. The sensor is shorted to chassis ground.
    D. The sensor is functioning correctly.

    Answer A is incorrect. An open sensor would be found by measuring between the two sensor leads and the meter reading out of limit (OL).

    Answer B is incorrect. This is not a test for voltage Wheel speed sensors can be checked for ac voltage with a voltmeter while spinning the wheel. .

    Answer C is correct. The sensor is shorted to chassis ground and must be replaced.

    Answer D is incorrect. The sensor is not functioning properly. This test is checking for a grounded sensor; if the sensor was functioning properly, an acceptable answer would be OL.

Section 6 Answer Keys and Explanations — Brakes (A5)

35. Technician A says the coil spring wrapped around a section of steel brake line is to help dissipate heat. Technician B says the spring wrapped around a section of steel brake line is to protect the line from abrasion wear. Who is correct?

    A. A
    B. B
    C. Both A and B
    D. Neither A nor B

    TASK A.2.4

    Answer A is incorrect. The wrap is not for heat dissipation.

    Answer B is correct. Only Technician B is correct. The wrap is for protection.

    Answer C is incorrect. Only Technician B is correct.

    Answer D is incorrect. Technician B is correct.

36. Technician A says the RWAL system may be deactivated on some vehicles when they are put in four-wheel drive mode. Technician B says some G force sensors are checked by measuring resistance while tilting the sensor. Who is correct?

    A. A
    B. B
    C. Both A and B
    D. Neither A nor B

    TASK F.1

    Answer A is incorrect. Technician B is also correct.

    Answer B is incorrect. Technician A is also correct.

    Answer C is correct. Both Technicians are correct. RWAL is deactivated on some vehicles because 4-wheel drive operation can produce a small amount of wheel slippage. G force sensors can be checked while tilting the sensor to ensure the sensor can measure changes in position.

    Answer D is incorrect. Both Technicians are correct.

Section 6 Answer Keys and Explanations

Brakes (A5)

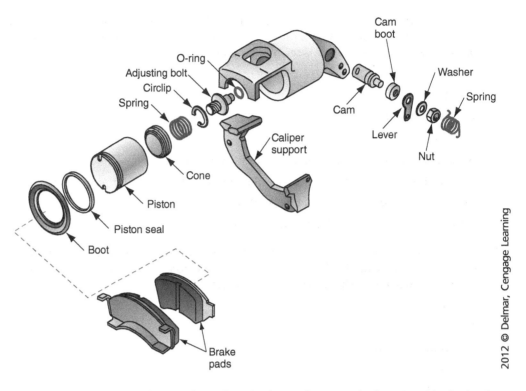

TASK E.3

37. Refer to the illustration above. The parking brake works correctly; however, the brake does not operate when the service brake pedal is pushed. Technician A says the problem could be a stuck adjuster. Technician B says the problem could be a seized cam. Who is correct?

   A. A
   B. B
   C. Both A and B
   D. Neither A nor B

Answer A is incorrect. A stuck adjuster would cause the parking brake to be out of adjustment.

Answer B is incorrect. A seized cam would prevent the parking brake from operating.

Answer C is incorrect. Neither Technician is correct.

Answer D is correct. Neither Technician is correct. If the service brake does not work and the parking brake does, there is a problem in the hydraulic circuit.

38. Rotor runout is checked with a dial indicator and found to be out of specification. The rotor is removed, rotated 180 degrees, and reinstalled. The runout is checked again and is still out of specification. The excessive runout is in the same location on the rotor. Which of the following is most likely indicated?

    TASK C.8

    A. The measurement is being performed incorrectly.
    B. The dial indicator is faulty.
    C. The runout is in the hub.
    D. The runout is in the rotor.

    Answer A is incorrect. The technicain is attempting to match mount (index) the rotor. This is a correct procedure.

    Answer B is incorrect. There is no reason to suspect the dial indicator is faulty.

    Answer C is incorrect. If the runout moved with the rotor, the runout is not in the hub.

    Answer D is correct. The runout moved with the rotor, indicating that the runout is in the rotor.

39. Which of the following would be the most likely cause of an ABS code?

    TASK F.4

    A. A grounded pad wear sensor
    B. Air in the brake fluid
    C. A cracked tone ring
    D. A malfunctioning proportioning valve

    Answer A is incorrect. A grounded pad wear sensor can cause the brake warning light to be illuminated but will not cause an ABS code.

    Answer B is incorrect. Air in the brake fluid can cause a spongy pedal but will not cause an ABS code.

    Answer C is correct. A cracked tone ring can cause the ABS to receive faulty wheel speed sensor information and set an ABS trouble code.

    Answer D is incorrect. A malfunctioning proportioning valve may cause the back brakes to fail to apply; however, it is very unlikely that it would cause an ABS code.

40. Brake fluid is leaking from the master cylinder reservoir cap. The seal is swollen and deformed. Which of the following could be the cause?

    TASK A.1.5

    A. Brake fluid contaminated with water
    B. Brake fluid contaminated with oil
    C. Brake fluid contaminated with metals
    D. Brake fluid has been overheated

    Answer A is incorrect. Fluid contaminated with water can be identified with test strips.

    Answer B is correct. Swollen rubber components are an indication of oil contamination.

    Answer C is incorrect. Metal contamination can be identified with test strips.

    Answer D is incorrect. Overheated brake fluid will be dark.

TASK C.9

41. Rotor runout is checked with a dial indicator and found to be out of specification. The rotor is removed, rotated 180 degrees, and reinstalled. The runout is checked again and found to be within specification. Which of the following is indicated?

   A. The rotor should be replaced.
   B. The hub should be replaced.
   C. The brake can be assembled in this position.
   D. The rotor should be turned 90 degrees and installed.

   Answer A is incorrect. Tthe rotor does not need to be replaced if the high and low spots can be match mounted so total runout is within specification.

   Answer B is incorrect. The hub does not need to be replaced if the high and low spots can be match mounted so total runout is within specification.

   Answer C is correct. If the high and low spots can be match mounted so total runout is within specification, the parts can be used. The brake can be assembled in this position, since runout is within specification.

   Answer D is incorrect. The rotor should be installed where the total runout is within specification.

TASK B.4

42. Which of the following is the hardest to remove from a drum with a drum brake lathe?

   A. Bell mouth
   B. Out of round
   C. Hard spots
   D. Grooving

   Answer A is incorrect. Bell mouth can be cut out of a drum.

   Answer B is incorrect. Out of round drums can be cut back to true on a drum lathe.

   Answer C is correct. The hardest to remove is hard spots. The cutting bit tends to raise and jump over the hard spot.

   Answer D is incorrect. Grooved drums can have the grooves machined out as long as the maximum drum dimension is not exceeded.

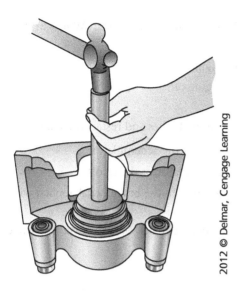

43. Technician A says the pin bushings are being installed in the illustration above. Technician B says the boot is being removed in the above illustration. Who is correct?

    A. A
    B. B
    C. Both A and B
    D. Neither A nor B

    TASK C.11

    Answer A is incorrect. Pin bushings are driven in and out with a punch.

    Answer B is incorrect. The boot is being installed.

    Answer C is incorrect. Neither Technicain is correct.

    Answer D is correct. Neither Technicain is correct. The boot is being installed.

44. When replacing the master cylinder, which of the following is true?

    A. The booster should also be replaced.
    B. The disc brake calipers should also be replaced.
    C. The drum brake wheel cylinders must also be replaced.
    D. The master cylinder should be bench bled prior to installation.

    TASK A.1.6

    Answer A is incorrect. The booster and master cylinder can be replaced separately.

    Answer B is incorrect. The master cylinder and calipers are replaced separately.

    Answer C is incorrect. The master cylinder and wheel cylinders can be replaced separately.

    Answer D is correct. The master cylinder should be bench bled prior to installing on the vehicle.

TASK D.4

45. A technician is testing the accumulator on a hydro-boost brake booster. Which of the following is a correct test procedure?

   A. Apply the brakes and hold, then start the engine.
   B. Apply the brakes and hold, then shut the engine off.
   C. Turn the engine off and apply the brakes.
   D. Apply the brakes, turn the engine off, release the brakes and restart the engine.

   Answer A is incorrect. This would test the booster not the accumulator.

   Answer B is incorrect. This would not test the accumulator or the booster.

   Answer C is correct. The engine should be shut off then the brake pedal applied while the technician feels for assist.

   Answer D is incorrect. This would not perform a definitive test on any portion of the brake system.

SECTION 7

Appendices

PREPARATION EXAM ANSWER SHEET FORMS

ANSWER SHEET

1. _____	21. _____	41. _____
2. _____	22. _____	42. _____
3. _____	23. _____	43. _____
4. _____	24. _____	44. _____
5. _____	25. _____	45. _____
6. _____	26. _____	
7. _____	27. _____	
8. _____	28. _____	
9. _____	29. _____	
10. _____	30. _____	
11. _____	31. _____	
12. _____	32. _____	
13. _____	33. _____	
14. _____	34. _____	
15. _____	35. _____	
16. _____	36. _____	
17. _____	37. _____	
18. _____	38. _____	
19. _____	39. _____	
20. _____	40. _____	

Delmar, Cengage Learning ASE Test Preparation

ANSWER SHEET

1. _____
2. _____
3. _____
4. _____
5. _____
6. _____
7. _____
8. _____
9. _____
10. _____
11. _____
12. _____
13. _____
14. _____
15. _____
16. _____
17. _____
18. _____
19. _____
20. _____
21. _____
22. _____
23. _____
24. _____
25. _____
26. _____
27. _____
28. _____
29. _____
30. _____
31. _____
32. _____
33. _____
34. _____
35. _____
36. _____
37. _____
38. _____
39. _____
40. _____
41. _____
42. _____
43. _____
44. _____
45. _____

ANSWER SHEET

1. _____
2. _____
3. _____
4. _____
5. _____
6. _____
7. _____
8. _____
9. _____
10. _____
11. _____
12. _____
13. _____
14. _____
15. _____
16. _____
17. _____
18. _____
19. _____
20. _____
21. _____
22. _____
23. _____
24. _____
25. _____
26. _____
27. _____
28. _____
29. _____
30. _____
31. _____
32. _____
33. _____
34. _____
35. _____
36. _____
37. _____
38. _____
39. _____
40. _____
41. _____
42. _____
43. _____
44. _____
45. _____

ANSWER SHEET

1. _____	21. _____	41. _____
2. _____	22. _____	42. _____
3. _____	23. _____	43. _____
4. _____	24. _____	44. _____
5. _____	25. _____	45. _____
6. _____	26. _____	
7. _____	27. _____	
8. _____	28. _____	
9. _____	29. _____	
10. _____	30. _____	
11. _____	31. _____	
12. _____	32. _____	
13. _____	33. _____	
14. _____	34. _____	
15. _____	35. _____	
16. _____	36. _____	
17. _____	37. _____	
18. _____	38. _____	
19. _____	39. _____	
20. _____	40. _____	

ANSWER SHEET

1. _____   21. _____   41. _____
2. _____   22. _____   42. _____
3. _____   23. _____   43. _____
4. _____   24. _____   44. _____
5. _____   25. _____   45. _____
6. _____   26. _____
7. _____   27. _____
8. _____   28. _____
9. _____   29. _____
10. _____   30. _____
11. _____   31. _____
12. _____   32. _____
13. _____   33. _____
14. _____   34. _____
15. _____   35. _____
16. _____   36. _____
17. _____   37. _____
18. _____   38. _____
19. _____   39. _____
20. _____   40. _____

ANSWER SHEET

1. _____
2. _____
3. _____
4. _____
5. _____
6. _____
7. _____
8. _____
9. _____
10. _____
11. _____
12. _____
13. _____
14. _____
15. _____
16. _____
17. _____
18. _____
19. _____
20. _____
21. _____
22. _____
23. _____
24. _____
25. _____
26. _____
27. _____
28. _____
29. _____
30. _____
31. _____
32. _____
33. _____
34. _____
35. _____
36. _____
37. _____
38. _____
39. _____
40. _____
41. _____
42. _____
43. _____
44. _____
45. _____

Glossary

Acetone A highly flammable liquid sometimes used to clean parts.

Aeration To expose to the air or mix with air, as with a liquid; to charge a liquid with gas.

Accumulator In ABS brake systems, a chamber that stores fluid. In hydro-boost brake systems, a charged chamber used to assist braking if a failure occurs to the power steering system or if the engine stalls.

Additive Anything that is added with the intention of improving a certain characteristic of the material or fluid.

Aerate To whip air into a fluid.

Air gap A small space between two parts.

Air shock A shock operating on the principles of air pressure; may also have a hydraulic section.

Alternating current (ac) An electric current whose polarity is constantly changing from positive to negative.

Antilock brake system (ABS) A computerized brake system that prevents wheel lockup by releasing pressure at the brake when the lockup is occurring.

Asbestos Heat resistant material once used extensively in brake linings and clutch discs. Inhaling asbestos dust is harmful and may cause cancer.

ASE Acronym for Automotive Service Excellence, a trademark of National Institute for Automotive Service Excellence.

Axis of rotation The center line around which a gear or part revolves.

Backing plate The part of a drum brake which the shoes and hardware mount to.

Battery A device for storing electrical energy in chemical form.

Bell mouth A term used to describe the condition of a drum when the drum has a greater diameter at the outer edge then the inner edge.

Bench bleeding A procedure for bleeding air from a master cylinder before installing it in a vehicle.

Bleeder screw A small valve-like screw located on each wheel cylinder and master cylinder used to bleed air from the brake system.

Bleeding The act of removing air from a hydraulic brake system.

Boiling point The temperature at which a substance, such as a liquid, begins to boil.

Booster A device which aids the driver in pushing the brake pedal.

Boot A flexible rubber or plastic cover over the end of wheel cylinders or master cylinder to keep out water and other foreign matter.

Bore May refer to the cylinder itself or to the diameter of the cylinder.

Brake A system used to slow or stop a vehicle; to slow or stop a vehicle.

Brake adjusting gauge tool A tool used on drum brakes to preset the brake shoes to the drum diameter.

Brake fade Loss of braking force due to heat buildup.

Brake fluid The liquid used in the brake system to transit force hydraulically.

Brake light Red lamps at the rear of a vehicle to warn others that a braking action is taking place.

Brake light switch A switch found on the brake linkage or in the hydraulic system that activates the brake lights.

Brake line A small diameter, rigid steel tube that connects the brake system to the brake hose which, in turn, connects to the wheel cylinders.

Brake pads A friction material applied to the disc by the caliper to slow or stop a vehicle.

Brake system The system in a vehicle that is used to slow or stop a vehicle.

Brake warning light An instrument panel lamp that warns of a brake system function or a malfunction.

Brinelling A condition in which a bearing or race has a series of dents or grooves worn into the surface.

Burnishing The procedure to break in a new set of brake pads.

Caliper The device used to apply force to the disc brake pads during braking. They can be sliding or fixed.

Check valve A valve, usually spring loaded, that allows the passage of fluid or vapor in one direction but not the other.

Circuit A complete path for an electric current; a compete path for a fluid system.

Coefficient of friction A measurement of the amount of friction developed between two objects in physical contact when one of the objects is drawn across the other.

Combination valve A valve in the brake system which combines the metering, pressure differential, and proportioning valve.

Compression Applying pressure to a spring or fluid.

Compressor Mechanical device that increases pressure within a circuit.

Compression fittings A type of fitting used to connect two lines.

Concentric grinding Grinding of the brake lining so it matches the brake drum.

Crocus cloth A fine abrasive cloth used for polishing or cleaning metal surfaces.

Cruise control A system that allows a vehicle to maintain a preset speed though the driver's foot is not on the accelerator pedal.

Dampen To slow or reduce oscillations or movement.

Denatured alcohol Ethyl alcohol used to clean brake parts.

Detergent additive An additive that helps keep metal surfaces clean and prevents deposits. These additives suspend particles of carbon and oxidized oil in the oil.

Diagnostic flow chart A chart that provides a systematic approach to the electrical system and component troubleshooting and repair. They are found in service manuals and are vehicle make and model specific.

Dial caliper A measuring instrument capable of taking inside, outside, depth, and step measurements.

Dial indicator A measuring tool used to determine end play motion.

Diaphragm A rubber seal located in the top of the master cylinder to keep fluid in and air out.

Digital Volt/Ohmmeter (DVOM) A test meter recommended for use on solid state circuits.

DOT An abbreviation for the Department of Transportation.

Drum That part of a brake that rotates with the wheel and that the brake shoes press against to slow or stop the vehicle.

Eccentric grinding Grinding portions of the brake lining to a different contour than that of the brake drum.

Electronic control unit A digital computer that controls engine and transmission functions.

Electronic stability control system (ESC) A computer controlled system found on automobiles to help the vehicle maintain a straight ahead position; it may reduce power to the wheels and/or apply individual brakes.

Equalizer A device in the brake cable system that prevents one side from being applied before the other.

Fault code A code that is recorded into the computer's memory.

Flare To spread gradually outward in a bell shape.

Floating caliper A disc brake caliper in which the caliper is mounted on pins or a slide to allow for some lateral movement.

Foot-pound (ft lb) An English unit of measurement for torque. One foot-pound is the torque obtained by a force of one pound applied to a foot long wrench handle.

Fretting Bearing damage as a result of vibration. The bearing outer race can pick up the machining pattern.

Galled bearing Bearing surface damage caused by overheating, lack of lubrication, or improper lubrication.

Glycol A term used for ethylene glycol, an antifreeze solution.

Grease Lubricant containing a mixture of oil, soap thickeners, and other ingredients.

Hazardous materials Any substance that is flammable, explosive, or is known to produce adverse health effects in people or the environment.

Heat A form of energy.

Heat checking Small surface cracks found in brake drums or brake rotors caused by excessive heat.

Hone To enlarge, smooth, and clean a cylinder bore using an abrasive stone.

Hub The center part of a wheel, gear, or bearing.

Housekeeping A routine of cleaning and other practices to insure a safe and healthy workplace environment.

Hydraulic booster A type of brake power booster that uses hydraulic pressure from the power steering system pump.

Hydro boost A power brake booster that utilizes hydraulic pressure from the power steering system pump.

Hygroscopic The ability to absorb moisture.

Ignition switch A multi-position master switch, usually key operated, in a vehicle.

Installation templates Drawings supplied by some vehicle manufacturers to allow the technician to correctly install the accessory. The templates available can be used to check clearances or to ease installation.

Integral ABS A self contained antilock brake system, which replaces the master cylinder and booster with a hydraulic modulator and high pressure accumulator.

Lamp A device used to convert chemical energy into radiant energy, usually visible light.

Lateral runout A rotating member that has excessive variations in the amount of sideways wobble when turning.

Leaf spring A rear vehicle suspension system component consisting of one or more flat leaves of steel with graduated lengths.

Lining A friction material attached to a brake part used to slow or stop a vehicle.

Linkage A system of rods and levers used to transmit motion or force.

Machine To remove metal in a grinding or cutting operation.

Maintenance manual A publication containing routine maintenance procedures and intervals for vehicle components and systems.

Manifold A device used to hold two or more instruments for testing purposes; a device used to channel the air/fuel mixture into an engine; a device used to channel exhaust vapors out of the engine.

Manifold vacuum The vacuum created by the motion of the pistons in the engine.

Master cylinder The part of the brake system which the driver operates by means of the brake pedal to send hydraulic pressure to the brakes.

Metering valve A valve in the brake system reduces pressure to the front brakes until the rear brakes apply.

Micrometer A precision measuring tool.

Mineral spirits A cleaning solvent.

Modulator A device that regulates hydraulic pressure.

Nitrogen A high pressure odorless, colorless, tasteless gas that is often used to pressurize a system for leak testing.

Non-integral ABS An antilock system which uses the existing master cylinder and booster, but has ABS components added to the brake system.

Normally closed A term that refers to a switch or valve in its normal position, closed.

Normally open A term that refers to a switch or valve in its normal position, open.

OEM Acronym for original equipment manufacturer.

Open circuit A circuit in which there is a break in continuity.

Pad wear indicator A mechanical or electrical warning device on the disc brake pad that warns of the need for pad replacement.

Parallelism Thickness variation in a brake rotor.

Parking brake A mechanically applied brake used to prevent a parked vehicle's movement.

Pascal's Law When pressure is exerted on a confined liquid, the pressure is transmitted undiminished.

Piston A round caliper component in a disc brake; an aluminum or sintered iron component of a drum brake inside a wheel cylinder; the valve-like rod in a master cylinder.

Pitting Surface irregularities resulting from corrosion.

Porosity A condition where fluids or gases can pass through the pores of a particular material.

Pounds per square inch (psi) A unit of English measure for pressure.

Power steering A steering system that used hydraulic pressure to increase the torque applied to the steering wheel.

Pressure differential valve A valve in the brake system which will turn on a light on the dash when the two circuits in the brake system have unequal pressure.

Pressure bleeder A device used to facilitate the removal of air from a brake system.

Pressure gauge A gauge which indicates pressure.

PRNDL lever The lever the driver moves to change gear positions in an automatic transmission.

Proportioning valve A valve in the brake system which reduces pressure to the rear brakes to help prevent brake lockup.

Pushrod A rod that transmits the movement and force of the wheel cylinder piston to the brake shoe.

Quick take-up master cylinder A master cylinder that supplies a large amount of fluid, under low pressure, during the first part of the brake application.

RBWL Red brake warning light on the dash.

Recall bulletin A bulletin that pertains to special situations that involve service work or replacement of components in connection with a recall notice.

Return springs Springs which pull the brake shoes away from the drum when the brakes are released.

Relay An electro-mechanical switch.

Reluctor ring A wheel that triggers or pluses a magnetic field.

Residual pressure Remaining or leftover pressure.

Rotor A disc shaped brake component that rotates with the wheel.

Runout Deviation or wobble of a shaft or wheel as it rotates. Measured with a dial indicator.

Scan tool An electronic equipment device designed to communicate with vehicle on-board computer systems. Trouble codes and data can be extracted using a scan tool.

Scoring A condition of deep scratches in a brake drum or disc brake rotor surface.

Shoe The lining and its steel backing of a drum brake.

Short circuit The intentional or unintentional grounding of an electrical circuit.

Silicone A group of organic compounds based on the element silicon (Si).

Sliding caliper A disc brake caliper that has piston(s) on one side of the caliper only.

Solenoid An electro-mechanical device used to impart a push-pull motion.

Specifications Technical data usually supplied by the vehicle manufacturer.

Spindle A shaft or axle on which a wheel hub or bearing rides.

Stoplight switch An electrical switch on the brake linkage or in the hydraulic system used to illuminate the brake lights when the brake pedal is depressed.

Swage To reduce or taper.

Torque A turning or twisting force.

Torque steer An outside influence of a front-wheel drive vehicle, such as uneven front tire-tread wear that causes the steering wheel to turn right or left during hard acceleration.

Toxicity A statement of how poisonous a substance is.

Traction control system (TCS) A computer controlled system which will reduce engine torque and or apply a brake to reduce wheel spin under acceleration.

Vacuum booster A power brake activation system that uses a vacuum signal on one side of a diaphragm to amplify braking effort.

Vacuum gauge A device used to measure vacuum.

Vacuum hose A small diameter rubber, plastic, or nylon tube used to transmit a vacuum signal.

Vacuum motor A diaphragm or motor-like device actuated by a vacuum.

Vacuum Pressure values below atmospheric pressure.

Vehicle Retarder An engine or driveline brake.

Vapor A gas.

Warning light A light, usually on the dash, to warn of a problem.

Wear light The manufacturers' specifications as to the durability of a part in terms of serviceability.

Wheel cylinder The device used to transfer hydraulic pressure in the brake system to the brake shoes.

Wheel lockup A condition that exists when the tire stops turning against the pavement and begins to slide.

Wheel sensor Magnetic speed sensor used to measure wheel speed in antilock brake systems.

Wiring harness The major assembly or a sub-assembly of a vehicle's wiring system.

Notes

Notes

Notes

Notes

Notes

Notes

Notes

Notes

Notes

Notes

Notes